# Edge Intelligence and Analytics for Internet of Things

Edge computing and intelligence is a new technical discipline that researches and develops theories, methods, technologies, and application systems for enabling fast and efficient processing and intelligent decision-making using the big data generated from the Internet of Things (IoT) system. The edge computing and intelligence–enabled IoT system helps in alleviating the resource limitations of IoT devices and facilitates large-scale deployment of IoT systems for different applications such as smart cities, smart grids, smart homes, smart healthcare systems, smart video surveillance, and e-healthcare.

Due to the deployment of the IoT devices in the open environment and their interaction with other IoT devices through unstable and unreliable wireless channels, as well as the placement of edge nodes near IoT systems, IoT devices offer many chances for malicious attackers to compromise them through collusion attacks, denial-of-service attacks, and many other types of attacks. Analysis of the network traffic and IoT data at the edge nodes using the latest artificial intelligence techniques will address a number of issues, such as anomaly detection in IoT data, data poisoning attacks, and other security and privacy issues related to IoT systems. In this book, we will provide an overview of the fundamental concepts of edge computing, edge intelligence using advanced machine- and deep-learning models, the architecture of edge computing–based IoT systems, and edge computing–based cyber threat intelligence solutions for IoT systems.

# Edge Intelligence and Analytics for Internet of Things

Edited by Govind P. Gupta, Brij B. Gupta,
Kumkum Garg, and Kwok Tai Chui

CRC Press
Taylor & Francis Group
Boca Raton London New York

CRC Press is an imprint of the
Taylor & Francis Group, an **informa** business

Designed cover image: Shutterstock Image ID 2271259129

First edition published 2026
by CRC Press
2385 NW Executive Center Drive, Suite 320, Boca Raton FL 33431

and by CRC Press
4 Park Square, Milton Park, Abingdon, Oxon, OX14 4RN

*CRC Press is an imprint of Taylor & Francis Group, LLC*



ISBN: 978-1-032-82349-2 (hbk)
ISBN: 978-1-032-84289-9 (pbk)
ISBN: 978-1-003-51206-6 (ebk)

DOI: 10.1201/9781003512066

Typeset in Sabon
by Apex CoVantage, LLC

# Contents

# Preface

Edge computing and intelligence is a new technical discipline that develops fundamental concepts, methods, technologies, and application systems for enabling fast and efficient processing and intelligent decision-making using the big data generated from Internet of Things (IoT) systems. The edge computing and intelligence-enabled IoT system helps in alleviating the resource limitation of IoT devices and facilitates large-scale deployment of IoT systems for different applications such as smart cities, smart grids, smart homes, smart healthcare systems, smart video surveillance, and e-healthcare. Due to the deployment of the IoT devices in open environments and their interaction with other IoT devices through unstable and unreliable wireless channels, as well as placement of edge nodes near the IoT systems, IoT devices offer many chances for malicious attackers to compromise them through collusion attacks, denial-of-service attacks, and many other types of attacks. This book provides an overview of the fundamental concepts of edge computing, edge intelligence using advanced machine- and deep-learning models, architecture of edge computing–based IoT systems, and edge computing–based cyber threat intelligence solutions for IoT systems.

**Govind P. Gupta,**
**Brij B. Gupta,**
**Kumkum Garg,**
**Kwok Tai Chui**

# About the editors

**Govind P. Gupta** is currently working as an Associate Professor in the Dept. of IT, National Institute of Technology Raipur, India. He has received his PhD in computer science and engineering from the Indian Institute of Technology, Roorkee, India. He has more than 15 years of teaching and research experience. He has made practical and theoretical contributions in big data processing and analytics, Wireless Sensor Network (WSN), Internet of Things (IoT), and cyber security domains, and his research has a significant impact on big data analytics, WSN, IoT, information, and network security against cyberattacks. He has published more than 150 research papers in reputed peer-reviewed journals and conferences of international reputes. He has received several R&D projects from prestigious funding organizations such as SERB (MATRIX), SERB(CRG) and C3iHub (IIT Kanpur). He is an active reviewer of various journals such as *IEEE Internet of Things, IEEE Sensors Journal, IEEE Transactions on Green Communications, Elsevier Computer Networks, Ad-Hoc Networks, Computer Communications, and Springer Wireless Networks*. His current research interests include IoT, IoT security, Edge Intelligence, network security, big IoT data analytics, blockchain-based application development for IoT, and enterprise blockchain. Dr. Gupta was recently included in the Global Alper-Doger (AD) Scientific Index 2024 list. He was also included in the 2024, 2023, 2022, and 2021 Stanford University lists of "Worlds Top 2%" scientists. He is a professional member of the IEEE and ACM.

**Prof. Brij B. Gupta** is working as Director of International Center for AI and Cyber Security Research and Innovations, and Distinguished Professor with the Department of Computer Science and Information Engineering (CSIE), Asia University, Taiwan. In more than 20 years of his professional experience, he published over 600 papers in journals/conferences including 35 books and 12 Patents with over 35,000 citations. He has received numerous national and international awards including Canadian Commonwealth Scholarship (2009), Faculty Research Fellowship Award (2017), MeitY, GoI, IEEE GCCE outstanding and WIE paper awards and Best Faculty Award (2018 & 2019), respectively. Prof. Gupta was selected for Clarivate Web of Science Highly Cited Researchers in Computer Science (top 0.1% researchers in the world) consecutively in 2022 and 2023. He was also selected in Stanford University's ranking of the world's top 2% scientists consecutively from 2019 to 2024. He is also a visiting/adjunct professor with several universities worldwide. Additionally, he is an IEEE Senior Member (2017) and was also selected as 2021 Distinguished Lecturer in IEEE CTSoc. Prof. Gupta was also a member of the Board of Governors, IEEE Consumer Technology Society (2022–2024). Prof. Gupta also leads IJSWIS, IJSSCI, and IJCAC as Editor-in-Chief. He also now serves as lead editor of a book series with CRC and IET press. He has been TPC member in more than 150 international conferences, and has been Associate/Guest Editor of various journals and publications. His research interests include information security, cyber physical systems, cloud computing, blockchain technologies, intrusion detection, AI, social media and networking.

**Kumkum Garg** obtained her bachelor of engineering (BE) and master of engineering (ME) from IIT Roorkee, India, and PhD from Imperial College of Science and Technology, University of London, UK. Her research interests are in computer networks, mobile computing, and AI. She has taught at the prestigious IIT Roorkee for 35 years. She has served as director at Manipal University and as the Dean, Faculty of Informatics, Photonics, and Robotics at Bhartiya Skill Development University, Jaipur, India. Currently, she is Honorary Professor at The North Cap University, Gurugram, India. She was awarded the Apple Distinguished Educator (ADE) and the Institution of Engineers (I) award. She has a patent for her work on mobile agent development, from the Government of India. She has also authored a book titled *Mobile Computing—Theory & Practice*, published by Pearson Education. Dr. Garg is Senior Member of IEEE, Fellow of Institution of Engineers (I), and Life Member of various professional societies. She has more than 47 years of experience in teaching and research and has over 135 research papers/publications in international and national journals and conferences.

**Kwok Tai Chui** received a BE in electronic and communication engineering with a business intelligence minor and PhD in electronic engineering from City University of Hong Kong. He has industry experience as a senior data scientist in an IoT company. He is with the Department of Technology, School of Science and Technology, at Hong Kong Metropolitan University as an assistant professor. He has more than 90 research publications, including edited books, book chapters, journal papers, and conference papers. He has served in various editorial positions in ESCI/SCIE-listed journals, including managing editor of the *International Journal on Semantic Web and Information Systems*, topic editor of *Sensors*, and associate editor of the *International Journal of Energy Optimization and Engineering*. His research interests include computational intelligence, data science, energy monitoring and management, intelligent transportation, smart metering, healthcare, machine learning algorithms, and optimization.

# Contributors

**R. Aishwarya**
Department of CSE
Sathyabama Institute of Science and Technology, India

**Padmalochan Bera**
School of Electrical and Computer Sciences,
Indian Institute of Technology Bhubaneswar, India

**Anish Chhetry**
Department of Computer Science and Engineering
National Institute of Technology Mizoram, India

**Vishal Choudhary**
School of Computer Applications
Lovely Professional University, Punjab, India

**Arun Kumar Dey**
Department of Information Technology
National Institute of Technology, Raipur, C.G., India

**Govind P. Gupta**
Department of Information Technology
National Institute of Technology, Raipur, C.G., India

**K. Ishwarya**
Computing Technologies, College of Engineering and Technology
SRM Institute of Science and Technology, Kattankulathur, Chennai, India

**Shareeful Islam**
Faculty of Science and Engineering
ARU, United Kingdom

**Jomina John**
School of Computer Science Engineering and Information Systems
Vellore Institute of Technology, Vellore, Tamil Nadu, India

**John Singh K.**
School of Computer Science Engineering and Information Systems
Vellore Institute of Technology, Vellore, Tamil Nadu, India

**Rahul Kala**
National Institute of Technology Mizoram

**Komal**
School of Computer Applications
Lovely Professional University,
Punjab, India

**Seema Kumari**
School of Computer Applications
Lovely Professional University,
Punjab, India

**Kishor Malakar**
National Institute of Technology
Mizoram

**Vaibhav Malviya**
National Institute of Technology
Mizoram

**G Mathivanan**
Department of CSE
Sathyabama Institute of Science and Technology

**Neerav Pawar**
Software Engineering Department
LENS, LUT University, Lappeenranta, Finland, 53850

**Madhukrishna Priyadarsini**
National Institute of Technology
Raipur, India

**Shalini Puri**
School of Information, Security and Data Science
Manipal University Jaipur, India

**Joseph James S.**
Computing Technologies, College of Engineering and Technology
SRM Institute of Science and Technology, Kattankulathur, Chennai, India

**Tirath Prasad Sahu**
Department of Information Technology
National Institute of Technology,
Raipur, C.G., India

**Gauri Shankar**
Software Engineering Department
LENS, LUT University, Lappeenranta, Finland, 53850

**Anjana Sharma**
School of Computer Applications
Lovely Professional University,
Punjab, India

**Avnish Singh**
Department of Information Technology
National Institute of Technology,
Raipur, C.G., India

**Rakesh Tripathi**
Department of Information Technology
National Institute of Technology,
Raipur, C.G., India

**Md Raihan Uddin**
Software Engineering Department
LENS, LUT University, Lappeenranta, Finland, 53850

**Upendra**
Department of Information Technology
National Institute of Technology,
Raipur, C.G., India

**Gurudatta Verma**
Department of Information Technology
National Institute of Technology,
Raipur, C.G., India

Chapter 1

# Edge analytics methods and algorithms

*Arun Kumar Dey and Govind P. Gupta*

## 1.1 Introduction

The proliferation of the Internet of Things (IoT) has led to a rise in intelligent gadgets equipped with numerous sensors, actuators, ubiquitous computing devices, and controllers resulting in the generation of massive amounts of data [1]. An advantage of this setup is that much of IoT data can be transferred to a centralized server or cloud for further processing and analysis, enabling more complex computational tasks and enhancing decision-making capabilities across connected networks [1]. However, the growing adoption of IoT-based intelligent gadgets has created challenges in managing massive amounts of data called big data [2]. Cisco Systems Inc. anticipates that approximately 500 billion connected devices will be operational by 2030, potentially generating an unprecedented volume of big data [3]. In addition, the cost of big data transmission and storage in the cloud is also high. Consequently, to optimize performance and reduce latency, computational tasks are increasingly performed on the edge layer, which consists of edge devices such as smartwatches, smartphones, and IoT gateways, rather than relying solely on cloud-based processing [1]. The setup is called edge technology, and it incorporates three terms such as edge devices, analytics, and computing [1], as depicted in Figure 1.1.

a) *Definition and concept of edge analytics*: In the edge technology framework, edge computing refers to computational tasks executed at edge devices, whereas edge analytics refers to data analysis conducted on these edge devices [3]. In edge analytics, data are gathered by edge devices, which are inside or closer to the edge layer [4]. Once the data are gathered, edge analytics uses some analysis methods and algorithms, such as video analytics or machine learning (ML), to interpret them. Next, edge analytics notifies authorities or raises an alarm based on predefined actions [3]. For example, edge computing, that is, computation on both edge devices and edge analytics infrastructure in industrial environments, enables automated monitoring of critical equipment parameters through sensor networks. This system provides continuous surveillance and automatic intervention when measurements exceed safety limits, addressing

DOI: 10.1201/9781003512066-1

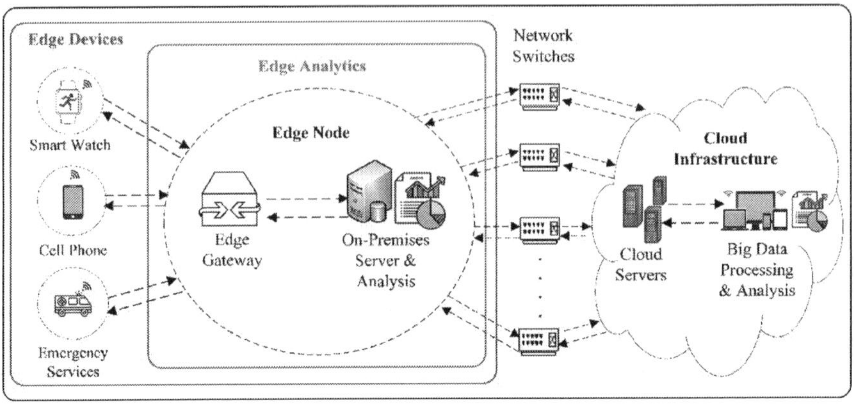


*Figure 1.1* Hierarchical framework of edge technology: interrelation among edge analytics, edge devices, and cloud computing.

the limitations and risks associated with human monitoring while preventing equipment damage and production failure.

*b) Motivation, benefits, and application of edge analytics*: The key motivation behind edge data analytics advancement is its capacity to minimize data transmission requirements to cloud servers and other computational resources for processing [4]. In addition, edge analytics reduces latency through proximity processing, optimized bandwidth utilization through selective data transmission, and enhanced privacy through localized data handling [4]. These benefits are complemented by improved system reliability, reduced communication costs, and flexible scalability options [1]. The application of edge analytics extends across multiple domains, encompassing healthcare diagnostics, industrial automation, retail intelligence, and financial security systems. This distributed edge analytical framework provides domain-specific solutions, facilitating real-time monitoring, predictive maintenance, and automated decision-making across various operational environments [1].

This chapter is structured into eight main segments. Section 2 presents architecture and working of edge analytics. Section 3 analyses various edge analytics categories. Section 4 edge analytics methods and algorithms are discussed. The chapter concludes with Section 5.

## 1.2 Architecture and working of edge analytics technology

*Architecture*: The architecture of edge technology incorporates three primary dimensions: edge devices, edge analytics, and edge computing [3]. Edge devices are further categorized into two parts: software and hardware.

A software-based edge device transforms data, formats output data, and so on, on an edge device such as an IoT-based intelligent device, whereas a hardware-based edge device is a physical intelligent device. Edge analytics refers to the various methodologies and algorithms used for data interpretation, while edge computing refers to computations performed by both edge devices and edge analytics. Hence, edge devices and analytics are encompassed by edge computing. Moreover, edge devices also process the outputs of edge analytics by adjusting, filtering, or compressing the data to make them suitable for cloud storage. Therefore, edge analytics operates as a functional component within edge devices, as shown in Figure 1.1.

Figure 1.1 demonstrates that edge analytics operates within the edge devices themselves rather than as an independent component. The hierarchical relationship depicted in Figure 1.1 can be formalized mathematically using set notation, as expressed in Eq. (1). This chapter therefore refers to edge devices analogously.

$$\{Edge\ Analytics\} \subseteq \{Edge\ Devices\} \subseteq \{Cloud\ Computing\}. \tag{1}$$

*Working of Edge Analytics*: In the working of edge analytics technology, edge devices like IoT-based intelligent gadgets with full-fledged operating systems and battery power continually capture data readings, with predefined upper and lower limits stored within the IoT-based intelligent devices [3]. Subsequently, the devices can execute edge intelligence to process data locally, utilizing an on-site server to perform operations and relay instructions to the edge gateway [4]. Edge analytics then uses algorithms within edge devices to analyse the data in real time, determining if any parameter exceeds the set thresholds. This process enables quick, localized decision-making. When parameters exceed their set thresholds, predefined actions are triggered within edge analytics, such as adjusting or shutting down the source generating the parameter (e.g., stopping fuel flow to minimize pressure) or notifying a human operator for assessment. These responses are programmed in advance.

Next, the processed data are subsequently transmitted to the cloud, either via a wireless connection or through intermediary network switches, facilitating efficient data management and informed decision-making [1]. Common protocols for transmitting data include Message Queuing Telemetry Transport, Hypertext Transfer Protocol Secure, JavaScript, Web Real-Time Communication, and Zigbee [4]. However, some types of data analysis require high computational methods and algorithms, such as artificial intelligence (AI–based models [3]. In these cases, model data training occurs in the cloud. Next, configurations, predicted results, or models are then sent to the edge devices by the cloud. After that, the edge analytics test the data using the predicted results. This localized processing minimizes latency, optimizes bandwidth usage, and enhances real-time decision-making. This enables edge analytics to effectively utilize complex algorithms directly within edge devices.

## 1.3 Types of edge analytics

Edge analytics can be categorized into three main types based on the data processed by edge devices (cloudlets) [3]: (1) Visual-based Unstructured Data Analytics, Textual-based Data Analytics, and Audio-based Data Analytics as depicted in Figure 1.2. Visual-based unstructured data analytics is further divided into Image-based Data Analytics and Video-based Data Analytics, each categorized into Descriptive Analytics (e.g., object recognition, activity detection) and Predictive Analytics (e.g., disease diagnosis, behaviour prediction). Textual-based data analytics is classified into Descriptive Text Analytics (e.g., summarization, sentiment analysis) and Predictive Text Analytics (e.g., intent detection, trend forecasting). Similarly, audio-based data, an emerging area in edge analytics, include Audio Signal Analytics (descriptive analysis of features like pitch and tone) and Audio Content Analytics (predictive tasks like emotion or intent detection). This structure highlights the evolution of edge analytics, with established focus areas in visual and textual data and growing potential in audio data analysis.

### *1.3.1 Edge analytics for visual-based unstructured data*

Visual-based unstructured data analytics, a key component of edge analytics as depicted in Figure 1.2, involve processing and analysing visual data directly on edge devices, enabling real-time insights while minimizing the need for data transmission to centralized servers. This category is further refined into two streams: Image-based and Video-based Data Analytics. Each stream encompasses both Descriptive Analytics—such as recognizing objects or detecting activities—and Predictive Analytics, including applications like diagnosing diseases and forecasting behaviour patterns. Such an approach is particularly beneficial for applications requiring immediate responses, such as surveillance and autonomous vehicles [5].

Li *et al.* [6] introduces the Age of Processed Information (AoPI) to optimize recognition accuracy and latency. Using a Lyapunov-based Block Coordinate Descent method, it efficiently manages edge server selection, resource allocation, and video configurations. This approach achieves up to a 10.94x improvement in AoPI, ensuring accurate and timely video analytics under dynamic network conditions. In addition, Fang *et al.* [7] developed an edge-assisted video analytics framework using a Bayesian optimization-based algorithm for context-aware offloading. It efficiently adapts to dynamic network and computational conditions, balancing inference accuracy and speed

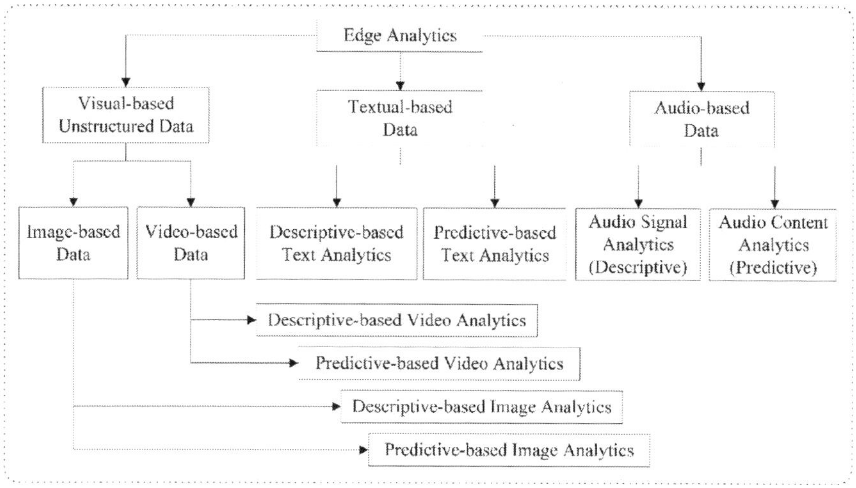


*Figure 1.2* Categories of edge analytics based on data types.

for real-time applications like surveillance and autonomous vehicles. Moreover, Qu *et al.* [8] introduces a learning-based framework for multi-drone edge orchestration in video analytics, employing supervised and reinforcement learning to optimize trajectories, video quality, and network protocols, ensuring efficient resource use and reliable performance for applications such as surveillance and smart farming.

Yuan *et al.* [9] proposes a split computing framework for visual data edge analytics, integrating scalable feature compression to enhance object detection and classification tasks. This method optimizes task accuracy under variable bitrates while reducing computational load on edge devices, making it efficient for bandwidth-constrained, real-time applications. Moreover, Xiao *et al.* [10] introduces an adaptive edge computing framework for video analytics that optimizes compression offloading and resource allocation. It employs a spatial attention-based compression strategy to preserve critical visual features, enhancing inference accuracy while minimizing bandwidth usage in real-time video processing. Table 1.1 presents the methods, characteristics, and limitations of five research studies on visual-based unstructured edge data analytics.

### 1.3.2 Textual data–based edge analytics

Textual data–based edge analytics can be divided into two categories: descriptive text analytics and predictive text analytics, each addressing specific analytical needs [3]. Descriptive text analytics involves summarizing and analysing textual datasets to identify patterns, trends, and relationships, offering a comprehensive understanding of the data. Predictive text analytics

*Table 1.1* Visual-based unstructured data for edge analytics methods and algorithms

| *Methods/Algorithms* | *Characteristics* | *Limitations* |
|---|---|---|
| AoPI Optimization [6] | Reduces delay and improves accuracy via resource allocation. | Limited adaptability to dynamic network variations. |
| Context-Aware Offloading Algorithm [7] | Optimizes accuracy and speed using Bayesian optimization. | High cost for real-time parameter tuning. |
| Multi-Drone Orchestration [8] | Adapts video settings and protocols using ML. | Coordination challenges in large drone networks. |
| Split Computing [9] | Enhances analytics with scalable feature compression. | Complex to implement across diverse hardware. |
| Collaborative Optimization model of Compression Rate and Resource Allocation [10] | Uses spatial-aware compression to optimize resources. | Balancing compression for varied visual data. |

uses ML methods and AI algorithms to develop models that predict future outcomes based on historical data. Predictive models are often trained in the cloud, and their inferences are deployed on edge devices to enable real-time analysis [3]. This approach reduces computational demands on edge devices but introduces a dependency on cloud resources. These approaches enable efficient and scalable processing of textual data at the edge, integrating real-time responsiveness with sophisticated analytical capabilities.

Symeonides *et al.* [11] introduces a query-driven descriptive analytics framework tailored for edge computing and IoT ecosystems. It introduces a declarative query model that enables efficient and scalable analysis of textual data streams, emphasizing latency reduction, query optimization, and real-time insights, suitable for diverse edge applications like transportation and healthcare. In addition, Cao *et al.* [12] proposes an edge computing platform for descriptive text analytics, leveraging mobile edge nodes to analyse real-time transit data streams. It introduces tasks like semantic annotation, temporal aggregation, and summary functions to uncover actionable insights, reducing latency and supporting decision-making in transit management. Moreover, Vasavi *et al.* [13] presents a predictive text-based edge analytics framework for vehicle health monitoring, using an ensemble of artificial neural network and k-nearest Neighbors algorithms (K-NN) to enhance fault prediction accuracy. This system processes sensor data in real time at the edge, reducing latency and ensuring timely risk alerts for improved vehicle safety and maintenance.

Nikolaou *et al.* [14] introduces a predictive edge analytics methodology using regression models for in-network processing. By caching and updating model parameters rather than transmitting raw data, the approach reduces communication overhead while maintaining prediction accuracy, optimizing efficiency for real-time IoT applications. Furthermore, Sarabia-Jácome *et al.* [15] presents an edge gateway architecture for predictive text–based analytics in healthcare scenarios, employing deep learning (DL) models for fall detection. By utilizing container-based virtualization, it improves resource management and reduces inference time compared to cloud-based systems, ensuring timely and efficient decision-making. Table 1.2 presents the methods, characteristics, and limitations of five research studies on text–based edge analytics.

### 1.3.3 Audio data–based edge analytics

Audio data–based edge analytics is a growing field that focuses on processing audio data at the edge to enable real-time insights [3]. It is categorized into audio signal analytics, which involves descriptive analysis of audio features such as pitch, tone, and frequency to understand and characterize the audio signals [16], and audio content analytics, which uses predictive techniques to infer higher-level information such as detecting emotions, identifying intent, or classifying speech patterns [16]. Together, these categories enhance the capability of edge devices to analyse audio data efficiently while reducing reliance on centralized processing.

Elliott *et al.* [16] explores audio-based edge analytics with a focus on environmental sound classification using lightweight DL models. It highlights the deployment of optimized neural networks for real-time audio analysis on resource-constrained edge devices, emphasizing reduced latency and efficient model performance for sound recognition tasks. In addition, Poroshenko *et al.* [17] provides a comprehensive overview of audio analytics system

*Table 1.2* Textual data–based edge analytics methods and algorithms

| *Methods/Algorithms* | *Characteristics* | *Limitations* |
|---|---|---|
| Query-Driven Descriptive Analytics [11] | Declarative model for real-time IoT data analysis. | Complex query optimization. |
| Real-Time Transit Data Descriptive Analytics [12] | Mobile edge nodes analyse transit data streams. | Limited scalability for large networks. |
| K-NN Predictive Analytics [13] | Ensemble model improves vehicle fault prediction. | High resource demand for execution. |
| In-Network Predictive Regression [14] | Cached models reduce data transmission overhead. | Accuracy trade-off with communication savings. |
| Edge Gateway Predictive Analytics [15] | Containerized DL models for fall detection. | Resource-intensive for deep learning models. |

*Table 1.3* Audio data–based edge analytics methods and algorithms

| *Methods/Algorithms* | *Characteristics* | *Limitations* |
|---|---|---|
| Lightweight convolutional neural network (CNN) for Environmental Sounds [16] | Optimized CNN for edge sound classification. | Limited scalability for diverse audio. |
| Audio Analytics Topologies [17] | Local processing for real-time audio events. | Hardware-dependent deployment. |
| Internet of Music Things (IoMT) [18] | Edge-based music analytics via crowdsensing. | Energy constraints in real-time use. |

topologies, emphasizing edge-based systems for processing sound events like alarms and intrusions. It highlights the efficiency of real-time audio signal analysis at the edge, reducing data transmission needs and enhancing responsiveness for critical applications such as security and monitoring. Moreover, Roy *et al.* [18] introduces an edge computing framework for music-based crowdsensing, leveraging audio analytics to process and classify sound data in real time. It highlights an energy-efficient and scalable architecture that integrates edge and cloud layers to improve responsiveness and minimize data transmission for IoT music applications. Table 1.3 presents the methods, characteristics, and limitations of three research studies on audio data–based edge analytics.

## 1.4 Advanced methods and algorithms for edge analytics

Researchers are continually striving to design advanced methods and algorithms for edge analytics that minimize computational demands. Edge technology enables the deployment of AI and ML models directly on edge devices, allowing edge nodes to learn from local data and adapt based on experience. This capability empowers edge devices to autonomously generate actions in response to events without requiring explicit programming. In this way, inference, being less resource-intensive, facilitates the execution of AI-based models on edge nodes, enabling real-time responses to user requests with minimal latency. Hence, these advancements highlight the potential of innovative methods and algorithms in enhancing the efficiency and responsiveness of edge analytics. The next portion of the section reviews several techniques that integrate AI-based models with edge analytics. Table 1.4 outlines the features and limitations of state-of-the-art edge analytics approaches utilizing AL and ML.

Microsoft introduced Azure IoT Edge [19], a fully managed service that integrates edge technology with ML by enabling IoT devices to perform data analytics and processing locally. It minimizes cloud dependency, enhances

*Table 1.4* Features and limitations of state-of-the-art edge analytics approaches

| *Techniques* | *Features* | *Limitations* |
|---|---|---|
| Microsoft Azure IoT Edge [19] | Local data processing with containerized workloads. | Cloud-dependent for advanced model updates. |
| AWS IoT Greengrass [20] | Real-time analytics with local and cloud integration. | Complex multi-device setup. |
| Cisco Edge Analytics Fabric [21] | Hyper distributed processing with reusable microservices. | Requires custom industrial configurations. |
| Dell Statistica [22] | Machine learning and edge analytics with visualization. | Limited scalability for high-frequency data. |
| HPE Edgeline Systems [23] | IT-OT integration with ruggedized edge hardware. | Compatibility issues with legacy systems. |
| IBM Watson IoT Edge Analytics [24] | Real-time monitoring and cognitive analytics. | Relies on cloud for model training. |
| Oracle Edge Analytics [25] | Event-driven real-time processing with SQL support. | High memory usage for complex events. |
| ThingWorx Analytics [26] | Predictive and anomaly detection with user-friendly tools. | Limited support outside ThingWorx platform. |
| SAP HANA Streaming Analytics [27] | Real-time streaming with ML integration. | Complex scalability configurations. |

response times to local events, and ensures reliability during offline periods. In addition, AWS Greengrass [20] extends cloud capabilities to edge devices, enabling local ML inference, secure communication, autonomous responses, and data preservation through advanced software and analytics features. Moreover, the Cisco Edge Analytics Fabric (EAF) system [21] provides a modular framework for IoT edge analytics, enabling real-time data processing and efficient data management across edge, fog, and cloud environments. It features reusable microservices, robust data transport, and integration with information technology (IT) systems for enhanced scalability and flexibility. By leveraging distributed processing, it minimizes latency, optimizes resource usage, and supports diverse IoT applications with comprehensive tools for development and operation. In addition, Dell Statistica [22], an advanced analytics platform originally developed by StatSoft, integrates machine learning, data visualization, and IoT edge scoring to enhance decision-making. Its distributed analytics architecture and reusable process templates provide scalable, flexible solutions across diverse sectors, including healthcare, manufacturing, and finance.

The HPE Edgeline [23] Converged Edge Systems integrate operational technology (OT) with IT to enable advanced edge analytics and secure data management. By combining rugged hardware, OT Link software, and Fortinet's security solutions, the system ensures real-time data processing, robust threat protection, and efficient operations in industrial environments, enhancing IoT and edge capabilities. In addition, the IBM Watson IoT Edge Analytics [24] is a cloud-managed platform that processes and transforms streamed data at the edge. It filters data, identifies critical conditions, enables local actions, and integrates with the cloud for advanced analytics, ML, and real-time monitoring. Moreover, Oracle Edge Analytics [25] leverages real-time event processing to analyse high-velocity data streams at the edge, enabling immediate insights and local actions. It integrates Java-based tools for scalable, low-latency event processing, supporting use cases like industrial automation, healthcare monitoring, and transportation logistics, while minimizing reliance on backend systems. Furthermore, ThingWorx Analytics [26] leverages AI and ML to provide real-time edge analytics, enabling predictive modelling, anomaly detection, and prescriptive insights for IoT applications. Integrated with the ThingWorx platform, it automates complex analytics processes, optimizes device performance, and enhances operational decision-making with minimal user expertise required. Additionally, SAP HANA Streaming Analytics [27] processes real-time data streams using event-driven logic for edge analytics, enabling immediate actions like alerts and notifications. It supports ML integration, pattern detection, and data transformation, optimizing IoT applications such as monitoring, predictive maintenance, and real-time decision-making. However, edge analytics faces several challenges with ML algorithms, including limited data availability, which leads to suboptimal learning, and the dynamic nature of real-time data, causing models to become outdated. Additionally, ML algorithms are computationally intensive, making them difficult to implement on resource-constrained edge nodes with limited power and processing capabilities [28].

## 1.5 Conclusion

Edge analytics has emerged as a transformative approach, enabling data processing closer to the source for faster insights and reduced cloud dependency. This chapter explored various methods for processing visual, textual, and audio data, highlighting the integration of AI and ML algorithms in enhancing the autonomy and efficiency of edge devices. In addition, edge analytics enhances cloud transmission and storage efficiency by filtering data locally, while enabling intelligent applications in industries like healthcare and manufacturing. However, despite its advancements, edge analytics still faces challenges such as dynamic data handling, computational limitations, and security concerns, necessitating further research. Future trends focus on federated learning for privacy, adaptive algorithms for dynamic data, and energy-efficient models to optimize edge device performance while ensuring security and scalability.

## References


[1] Andročec, D. (2023). Applications of edge analytics: A systematic review. *Acta Universitatis Sapientiae, Informatica*, *15*(2), 345–358.

[2] Nayak, S., Patgiri, R., Waikhom, L., & Ahmed, A. (2024). A review on edge analytics: Issues, challenges, opportunities, promises, future directions, and applications. *Digital Communications and Networks*, *10*(3), 783–804.

[3] Camhi, J. (2015). Former cisco CEO John Chambers predicts 500 billion connected devices by 2025. https://www.businessinsider.com/former-cisco-ceo-500-billion-connected-devices-by-2025-2015-11, [Online; accessed 9 November 2022].

[4] Jayashree, N., & Babu, B. S. (2022). Edge data analytics technologies and tools. In *Advances in Computers* (Vol. 127, pp. 209-236). Elsevier.

[5] Xu, R., Razavi, S., & Zheng, R. (2023). Edge video analytics: A survey on applications, systems and enabling techniques. *IEEE Communications Surveys & Tutorials*, *25*(4), 2951–2982. https://doi.org/10.1109/COMST.2023.3323091.

[6] Li, X., Zhang, S., Huang, Y., Ma, X., Wang, Z., & Luo, H. (2024, November). Towards timely video analytics services at the network edge. *IEEE Transactions on Mobile Computing*, *23*(11), 10443–10459. https://doi.org/10.1109/TMC.2024.3376769.

[7] Dai, P., Chao, Y., Wu, X., Liu, K., & Guo, S. (2024). Context-aware offloading for edge-assisted on-device video analytics through online learning approach. *IEEE Transactions on Mobile Computing*, *23*, 12761–12777.

[8] Qu, C., Singh, R., Esquivel-Morel, A., & Calyam, P. (2024). Learning-based multi-drone network edge orchestration for video analytics. *IEEE Transactions on Network and Service Management*, *21*, 6331–6348.

[9] Yuan, Z., Rawlekar, S., Garg, S., Erkip, E., & Wang, Y. (2024). Split computing with scalable feature compression for visual analytics on the edge. *IEEE Transactions on Multimedia*, *26*, 10121–10133.

[10] Xiao, W., Hao, Y., Liang, J., Hu, L., Alqahtani, S. A., & Chen, M. (2024). Adaptive compression offloading and resource allocation for edge vision computing. *IEEE Transactions on Cognitive Communications and Networking*, *10*, 2357–2369.

[11] Symeonides, M., Trihinas, D., Georgiou, Z., Pallis, G., & Dikaiakos, M. (2019, June). Query-driven descriptive analytics for IoT and edge computing. In *2019 IEEE International Conference on Cloud Engineering (IC2E)* (pp. 1–11). IEEE.

[12] Cao, H., Wachowicz, M., & Cha, S. (2017, December). Developing an edge computing platform for real-time descriptive analytics. In *2017 IEEE International Conference on Big Data (Big Data)* (pp. 4546–4554). IEEE.

[13] Vasavi, S., Aswarth, K., Pavan, T. S. D., & Gokhale, A. A. (2021). Predictive analytics as a service for vehicle health monitoring using edge computing and K-NN algorithm. *Materials Today: Proceedings*, *46*, 8645–8654.

[14] Nikolaou, S., Anagnostopoulos, C., & Pezaros, D. (2019, April). In-network predictive analytics in edge computing. In *2019 Wireless Days (WD)* (pp. 1–4). IEEE.

[15] Sarabia-Jácome, D., Lacalle, I., Palau, C. E., & Esteve, M. (2019, April). Efficient deployment of predictive analytics in edge gateways: Fall detection scenario. In *2019 IEEE 5th World Forum on Internet of Things (WF-IoT)* (pp. 41–46). IEEE.

[16] Elliott, D. (2020). *Efficient Edge Analytics: Addressing Cyber-Physical MASINT with Machine Learning on Audio at the Edge*. Florida Institute of Technology, Scholarship Repository @ Florida Tech.

[17] Poroshenko, A., Kovalenko, A., & Sedlaček, P. (2022, October). Organization of audio analytics systems topologies. In *2022 IEEE 9th International Conference on Problems of Infocommunications, Science and Technology (PIC S&T)* (pp. 517–521). IEEE.

[18] Roy, S., Sarkar, D., Hati, S., & De, D. (2018). Internet of music things: An edge computing paradigm for opportunistic crowdsensing. *The Journal of Supercomputing*, *74*, 6069–6101.

[19] Technologies, T. Iot edge analytics & machine learning for real-time device control. https://www.thinxtream.com/whitepapers/thinxtream-iot-edge-analytics-machine-learning-real-time-device-control-wp-003.pdf. (Accessed 27 November 2019).

[20] Zimine, D. Aws greengrass—The missing manual. https://cbusby.me/post/173205913172/aws-greengrass-the-missing-manual. (Accessed 16 January 2020).

[21] https://www.cisco.com/c/dam/en/us/products/collateral/analytics-automation-software/edge-analytics-fabric/eaf-whitepaper.pdf.

[22] Demidova, L., Ivkina, M., Zhdankina, E., Krylova, O., Sofyin, E., Reshetova, V., . . . Tyart, N. (2016). Software package statistica and educational process. In *SHS Web of Conferences* (Vol. 29, p. 02011). EDP Sciences.

[23] https://www.fortinet.com/content/dam/fortinet/assets/alliances/sb-hpe-edgeline-converged-edge-systems-and-ot-link.pdf.

[24] https://www.ibm.com/docs/en/mapms/1_cloud?topic=analyzing-data.

[25] https://www.oracle.com/technetwork/middleware/complex-event-processing/overview/edge-analytics-ds–2775436.pdf.

[26] https://econocap.com/wp-content/uploads/2018/01/ThingWorx-Analytics-Product-Brief-.pdf.

[27] https://help.sap.com/doc/2481ac53344f4976856569e34a93e6c3/2.0.04/en-US/streaming_master_guide.pdf.

[28] Aral, A., & Brandic, I. (2019). Consistency of the fittest: Towards dynamic staleness control for edge data analytics. In: Mencagli, G., et al. (Eds.), *Euro-Par 2018: Parallel Processing Workshops. Euro-Par 2018. Lecture Notes in Computer Science*, vol. 11339. Springer, Cham. https://doi.org/10.1007/978-3-030-10549-5_4

Chapter 2

# Introduction to edge computing and IoT

*Vishal Choudhary, Shalini Puri, Komal, and Anjana Sharma*

## 2.1 Introduction

By facilitating quicker, more effective, and secure data processing, the convergence of Edge Computing (EC) and Internet of Things (IoT) technologies is drastically altering the global digital environment. Connecting different objects, including sensors, cars, and household appliances, via the internet and remotely controlling them is the function and the goal of the IoT [1–10]. The IoT and EC are rapidly developing technologies that are transforming many industries by enhancing the capabilities, dependability, and efficiency of digital systems. This method decreases the latency, reduces bandwidth consumption, and boosts data privacy and security by restricting the transmission of sensitive data to the cloud. The things connected over the internet can collect, exchange, and act on data in real time. the connected devices disseminate vast amounts of information that can improve decision-making capabilities, help in automation, and create new insights in applications ranging from smart homes and healthcare to industrial automation and environment monitoring [11–15]. IoT devices generate enormous amounts of data, which can be challenging to handle and manage, particularly when relying on traditional cloud computing techniques. Sending all data to a centralized cloud for processing could result in delays, network congestion, and bandwidth expenses. EC replaces IoT as an auxiliary technology. It decreases the distance that data must travel by locating data processing and storage near the sources and devices that supply the data [16–19].

EC is perfect for applications that need quick reaction times, such as remote healthcare monitoring, industrial machinery management, and driverless cars, because it can drastically lower latency by moving computation to the network's edge [20–24]. Additionally, EC improves data security and privacy by storing private information on local networks rather than sending it across potentially unsafe internet channels. In addition to improved speed and security, EC can also help reduce operational costs by lowering the volume of data sent to the cloud, thereby conserving bandwidth and reducing cloud storage expenses [25]. Together, IoT and EC provide a powerful

DOI: 10.1201/9781003512066-2

framework for creating highly responsive, data-driven environments that adapt to real-time demands, supporting innovative applications in diverse sectors such as smart cities, logistics, agriculture, and beyond [26, 27]. Their combined influence promises a future where digital and physical worlds are seamlessly integrated, enabling smarter, more connected, and more efficient systems on a global scale.

The rest of the chapter is organized as follows. Section 2.2 presents the history of IoT including early conceptualization in the late 20th century, development of Radio-Frequency Identification (RFID) and sensor technologies in 1990, proliferation of wireless networks in 2000, smart home and consumer IoT in early 2010, Industrial IoT (IIoT) emergence in mid-2010, standardization efforts in late 2010, EC and artificial intelligence (AI) integration from 2010 to 2020, and current trends in 2020. Section 2.3 discusses the architectural details of IoT including components of IoT, along with key concepts and principles. Sections 2.4 and 2.5 discuss the emergence of EC and the benefits of EC, respectively. Section 2.6 explains the intersection of EC and IoT, and Section 2.7 describes the complementary strengths of IoT and EC. Section 2.8 discusses the role of EC in IoT applications. Furthermore, Section 2.9 addresses the challenges in integrating EC and IoT, and Section 2.10 provides the challenges and future directions. Section 2.11 concludes the chapter.

## 2.2 The history of IoT

The history of the IoT traces back to the late 20th century, although its conceptual roots are found earlier. The history of IoT reflects a journey of technological innovation and adoption, driven by the vision of a more interconnected and intelligent world.

### 2.2.1 *Early conceptualization in the late 20th century*

The late 20th century saw the emergence of the concept of networked devices interacting with one another without the need for human interaction. One of the first internet-connected devices was a modified Coke machine at Carnegie Mellon University in 1982, which could report its inventory and whether freshly filled beverages were cold [2].

### 2.2.2 *Development of RFID and sensor technologies in 1990*

The evolution of RFID and sensor technologies has played a pivotal role in shaping the landscape of modern connectivity and data collection. RFID has developed significantly, transforming from its early conceptualization to widespread implementation in various industries. Initially conceived as a means of uniquely identifying objects, RFID technology has matured to encompass a diverse range of applications, from inventory management in retail to access control in transportation systems. Similarly, sensor technologies have experienced remarkable advancements, driven by real-time data acquisition and analysis demand. From simple temperature and humidity sensors to sophisticated environmental and biometric sensors, the capabilities of sensor devices have expanded exponentially. These advancements have enhanced the accuracy and reliability of data collection and enabled the integration of sensor networks into IoT ecosystems, facilitating seamless communication and data exchange between interconnected devices. The way we engage with the physical world has been completely transformed by RFID and sensor technologies working together to provide previously unheard-of levels of automation, efficiency, and intelligence. The uses of these technologies will only grow more varied and significant as they develop further, influencing everything, from smart cities and healthcare to manufacturing and logistics [3].

### 2.2.3 *Proliferation of wireless networks in 2000*

The widespread adoption of wireless networks, notably Wi-Fi and cellular networks, has ushered in an era of seamless connectivity, providing the essential infrastructure required to interconnect devices across vast distances. With the proliferation of these wireless technologies, the limitations of traditional wired connections have been surpassed, paving the way for a dramatic expansion of IoT applications across diverse industries and environments. Wi-Fi networks, characterized by their high-speed data transmission capabilities and widespread availability, have enabled the integration of IoT devices into existing network infrastructures with unparalleled ease [4]. From smart homes and offices to retail spaces and industrial facilities, Wi-Fi connectivity has become the backbone of countless IoT deployments, facilitating real-time data exchange and remote device management [5].

### 2.2.4 *Smart home and consumer IoT in the early 2010s*

In the early 2010s, there was a significant surge in the adoption of consumer IoT devices, particularly within the burgeoning smart home sector. This era marked the emergence of a wide array of innovative products aimed at enhancing convenience, comfort, and security in residential environments. Among the most notable advancements were smart thermostats, lighting systems, and security cameras, which rapidly gained traction among homeowners seeking greater control and automation capabilities. Smart thermostats revolutionized home climate control by leveraging IoT technology to offer personalized heating and cooling solutions tailored to individual preferences and usage patterns. Equipped with sensors and intelligent algorithms, these devices not only optimized energy efficiency but also provided users with remote access via smartphone apps, allowing them to adjust temperature settings and monitor energy consumption from anywhere, at any time. Similarly, IoT-powered lighting systems offer unprecedented flexibility and convenience in home illumination, enabling users to customize lighting schedules, colors, and brightness levels to suit their preferences and moods. Smart lighting solutions, which turn everyday living spaces into immersive settings, offer easier control and better ambiance through features like voice control and connection with virtual assistants like Google Assistant and Amazon Alexa. Moreover, the proliferation of IoT-enabled security cameras revolutionized home surveillance and monitoring, empowering homeowners to keep a watchful eye on their property and loved ones remotely. Equipped with high-definition cameras, motion sensors, and two-way audio communication, these devices offered real-time alerts and live streaming capabilities, providing peace of mind and proactive security measures against potential threats or intrusions [6].

### *2.2.5 IIoT emergence in the mid-2010s*

The IIoT has emerged as a pivotal and expansive application domain, harnessing the capabilities of IoT technologies to drive transformative advancements in efficiency, productivity, and safety across a myriad of industrial sectors. This encompassing landscape includes a diverse array of applications spanning manufacturing, agriculture, healthcare, logistics, and beyond, each leveraging the power of connected devices and data-driven insights to optimize operations and enhance overall performance. In the realm of manufacturing, IIoT technologies have revolutionized traditional production processes by enabling real-time monitoring, predictive maintenance, and adaptive control systems. Smart sensors embedded within machinery and equipment collect and transmit valuable operational data, allowing for proactive maintenance interventions, predictive analytics, and continuous optimization of production workflows. This heightened level of automation and intelligence not only minimizes downtime and reduces maintenance costs but also enhances overall equipment effectiveness and product quality.

In agriculture, IIoT solutions are driving the modernization of farming practices and agribusiness operations, facilitating precision agriculture techniques and sustainable resource management. IoT-enabled sensors deployed across fields and livestock facilities monitor environmental conditions, soil moisture levels, crop health, and animal behavior, enabling farmers to make data-driven decisions and optimize resource allocation. From automated irrigation systems to smart pest management solutions, IIoT technologies empower agricultural stakeholders to maximize yields, minimize inputs, and mitigate risks associated with weather variability and environmental factors. Similarly, in healthcare, IIoT applications are revolutionizing patient care delivery, medical diagnostics, and facility management through the integration of smart medical devices, wearable sensors, and connected infrastructure.

IoT-enabled medical technologies, such as smart implants and remote patient monitoring systems, allow real-time data transfer and ongoing health monitoring, enabling healthcare professionals to provide proactive interventions and individualized treatment programs. Moreover, IoT-driven asset tracking and facility management solutions streamline operations, optimize resource utilization, and enhance patient safety within healthcare facilities. In the logistics sector, IIoT technologies are driving efficiency improvements and cost reductions through enhanced supply chain visibility, asset tracking, and predictive analytics. IoT-enabled sensors and tracking devices deployed across transportation networks, warehouses, and distribution centers monitor cargo conditions, location, and transit routes in real time, enabling proactive risk management, route optimization, and demand forecasting. This heightened level of visibility and control enhances supply chain resilience, reduces lead times, and improves customer satisfaction across diverse industries [7].

### 2.2.6 *Standardization efforts in the late 2010s*

Standardization efforts, spearheaded by renowned organizations such as the International Telecommunication Union (ITU), the Institute of Electrical and Electronics Engineers (IEEE), and the Internet Engineering Task Force (IETF) have been pivotal in confronting the interoperability and security hurdles encountered within IoT deployments. These collaborative endeavors seek to establish uniform frameworks, protocols, and specifications that ensure seamless integration and communication among disparate IoT devices and systems, irrespective of their manufacturer or underlying technology. The ITU, a specialized United Nations agency, plays a central role in developing global standards and recommendations for telecommunications and Information and Communication Technologies (ICTs). Through its dedicated Study Groups and expert committees, the ITU fosters consensus-driven processes to address key IoT standardization challenges, ranging from network protocols and data formats to security mechanisms and interoperability guidelines.

Similarly, the IEEE, renowned for its contributions to technological innovation and advancement, actively engages in IoT standardization efforts through its numerous working groups and technical committees. Leveraging its vast pool of expertise and resources, the IEEE facilitates the development of open, consensus-based standards that underpin the interoperability, reliability, and security of IoT ecosystems across diverse domains and applications. Furthermore, the IETF, a premier international standards organization, focuses on the development and promotion of internet standards and protocols essential for the functioning of the global internet. Recognizing the pivotal role of IoT in shaping the future of interconnected systems, the IETF actively engages in standardization activities related to IoT networking, security, and data management, fostering the creation of robust, scalable, and secure IoT architectures. By collaborating with industry stakeholders, academia, and government bodies, these leading standardization organizations endeavor to address the multifaceted challenges inherent in IoT deployments, ranging from interoperability issues and protocol fragmentation to security vulnerabilities and privacy concerns. Through the establishment of consensus-driven standards and best practices, they strive to foster innovation, accelerate market adoption, and ensure the long-term sustainability and interoperability of IoT ecosystems on a global scale [8].

### 2.2.7 EC and AI integration from 2010 to 2020

EC, heralded for its paradigm shift toward localized data processing, surged in prominence within IoT deployments. This transformative approach prioritizes processing data near its point of origin, diverging from the conventional reliance on centralized data centers. IoT architectures that use EC benefit greatly from lower latency, less bandwidth use, and improved data security and privacy. This development overcomes the intrinsic drawbacks of conventional cloud-based models and gives IoT systems more scalability, dependability, and efficiency. Additionally, incorporating AI technology, such as computer vision and machine learning (ML), has enhanced IoT systems' capacity for data analytics and decision-making to previously unheard-of levels. Real-time processing and analysis of sensory data streams is made possible by AI-driven algorithms that may be installed directly on IoT devices or at the network's edge. This allows for adaptive responses, anomaly identification, and predictive insights. ML models leverage historical data to identify patterns, trends, and correlations, empowering IoT systems to autonomously optimize operations, anticipate maintenance needs, and enhance user experiences.

Moreover, computer vision technologies enable IoT devices to interpret and understand visual data captured from cameras or sensors, facilitating applications such as object recognition, activity detection, and environmental monitoring. By harnessing the power of AI and computer vision at the edge, IoT systems gain the intelligence to perceive, comprehend, and act upon their surroundings in a context-aware manner. This fusion of EC and AI not only amplifies the analytical capabilities of IoT deployments but also unlocks new possibilities for innovative applications across diverse domains, including smart cities, autonomous vehicles, healthcare, and industrial automation. In essence, the convergence of EC and AI represents a paradigm shift in the landscape of IoT, propelling the evolution of intelligent, autonomous, and responsive systems. By leveraging localized data processing and AI-driven analytics, IoT deployments are poised to deliver transformative outcomes, driving efficiency gains, operational enhancements, and enriched user experiences in the interconnected world of tomorrow [9].

### 2.2.8 *Current trends in 2020*

In recent years, the evolution of IoT has been characterized by significant advancements across various fronts, including connectivity, EC, AI, and cybersecurity. These progressive developments have propelled IoT into new realms of innovation and application diversity, reshaping industries and revolutionizing how we interact with technology. The advent of cutting-edge connectivity solutions, exemplified by the rollout of fifth-generation (5G) networks, has unlocked unprecedented possibilities for IoT deployments. With its ultra-fast speeds, low latency, and massive device capacity, 5G technology has catalyzed the proliferation of IoT applications across diverse domains. From smart cities harnessing IoT sensors to optimize infrastructure and enhance urban services to connected vehicles leveraging real-time data for intelligent navigation and vehicle-to-vehicle communication, the transformative potential of 5G-enabled IoT ecosystems is boundless.

Moreover, the convergence of EC and IoT has ushered in a new era of distributed intelligence and real-time responsiveness. By processing data closer to its source, EC augments IoT deployments' efficiency, reliability, and scalability, enabling rapid decision-making, reduced latency, and enhanced privacy and security. This synergy between EC and IoT has unlocked novel use cases in healthcare wearables, where real-time health monitoring and personalized patient care are revolutionizing healthcare delivery, and industrial automation, where edge-powered analytics optimize production processes and ensure operational efficiency. Additionally, data analytics, predictive modeling, and decision-making capabilities have all been transformed by the incorporation of AI technology into IoT ecosystems. Predictive maintenance, anomaly detection, and intelligent automation across a range of applications are made possible by AI-powered IoT systems that use ML algorithms to extract meaningful insights from massive sensor data streams.

From smart home devices adapting to user preferences to industrial robots optimizing manufacturing workflows, AI-driven IoT solutions are redefining efficiency, productivity, and user experiences across industries. However, alongside these remarkable advancements, the importance of cybersecurity in IoT deployments cannot be overstated. As IoT devices become increasingly interconnected and pervasive, they also become prime targets for cyberattacks and data breaches. Robust cybersecurity measures, including encryption, authentication, access control, and regular software updates, are imperative to safeguard IoT ecosystems and protect sensitive data from malicious actors [10].

## 2.3 Architectural details of IoT

The IoT encapsulates a vast network of interconnected physical entities or things imbued with an array of sensors, software, and other cutting-edge technologies. These components synergize to enable seamless data collection and

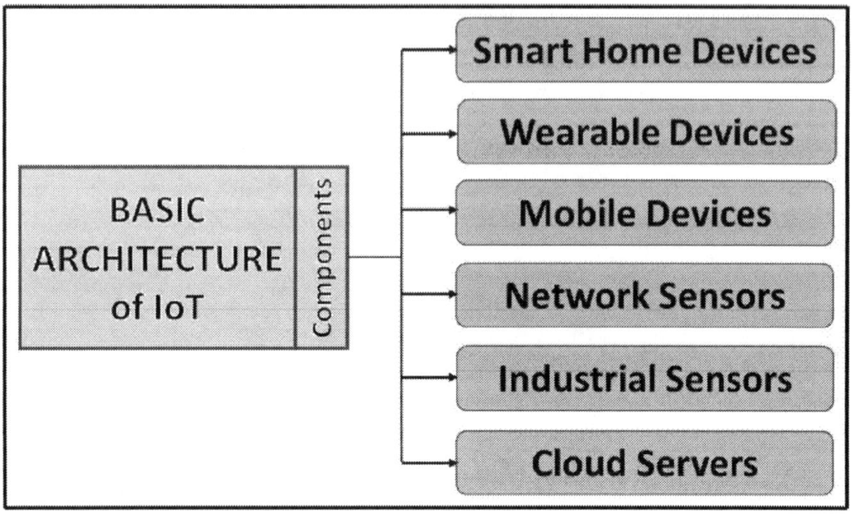


*Figure 2.1* Basic architecture of an IoT application.

exchange across the internet, heralding a new era of connectivity and intelligence [11]. Within this expansive ecosystem, the spectrum of IoT-enabled objects spans a diverse array of functionalities and applications. From the familiar terrain of everyday household items like smart thermostats, refrigerators, and wearable devices, to the sprawling landscapes of industrial machinery and critical infrastructure, the scope of IoT extends far and wide. These IoT-enabled entities serve as the digital conduits bridging the physical and virtual realms, facilitating real-time data acquisition, analysis, and decision-making processes. Whether optimizing energy consumption in smart homes, enhancing operational efficiency in industrial settings, or enabling predictive maintenance in transportation networks, IoT stands poised to revolutionize industries and transform the fabric of daily life. Figure 2.1 shows the basic architecture of an IoT application that includes smart home devices, wearable devices, mobile devices, network devices, industrial sensors, and cloud servers.

### 2.3.1 *Components of IoT*

The following are the main components of the IoT:

- Things/Devices: These are the physical objects equipped with sensors, actuators, and connectivity hardware that enable them to interact with the environment and exchange data.
- Sensors and Actuators: Sensors collect information from their surroundings by detecting various physical attributes like temperature, pressure,

light, and motion. Meanwhile, actuators empower devices to execute tasks in response to the data provided by sensors, such as toggling power, fine-tuning configurations, and more.

- Connectivity: Wi-Fi, Bluetooth, Zigbee, RFID, cellular networks (3G, 4G, 5G), Long Range Wide Area Network (LoRaWAN), and other communication technologies are used by IoT devices to connect to the internet or local networks.
- Data Processing and Analytics: Information gathered by IoT devices is processed, examined, and frequently kept in the cloud or at the network's edge. Decision-making, automation, and real-time insights are made possible by this data processing [12].
- User Interface: Platforms such as mobile applications, web interfaces, or command-line systems provide users with the capability to oversee and manage IoT devices from afar. This includes tasks like adjusting settings, receiving notifications, and accessing data analytics.
- Security and Privacy Mechanisms: To safeguard IoT ecosystems, it is essential to put security mechanisms like encryption, authentication, access control, and regular software upgrades into place, given the sensitive nature of IoT data and the possible threats posed by connected devices [13].

### 2.3.2 *Key concepts and principles*

- Interoperability: Ensuring interoperability among IoT devices and systems, regardless of their manufacturer or underlying technology, is vital for seamless communication and operation [14].
- Scalability: The scalability of IoT systems is paramount in accommodating an expanding array of devices and managing escalating volumes of data with efficiency. As the network of interconnected devices continues to grow, IoT systems need to possess the flexibility and robustness to seamlessly integrate new devices without compromising performance. Scalability enables IoT infrastructures to adapt to dynamic demands, ensuring optimal resource utilization and enhancing overall system reliability. By employing scalable architectures and innovative solutions, IoT ecosystems can effectively manage the proliferation of devices and data, thereby maximizing their potential impact across various domains and industries.
- Reliability: IoT systems should be robust and resilient, capable of operating in diverse environments and under varying conditions without compromising performance.
- Privacy and Ethics: Gaining consent, anonymizing sensitive data, and following rules and guidelines to respect user privacy and ethical issues when gathering, storing, and using IoT data are important [15].
- EC: In IoT applications, processing data closer to the source, or EC, has several benefits. Since there is no need to send data back and forth to a

central server, latency is reduced by doing computations close to the network's edge, where data are generated. IoT systems become more responsive, nimble, and able to make decisions in real time because of this latency reduction. By lowering the amount of data that must be sent over the network, EC also aids in bandwidth conservation, which is especially advantageous in settings with poor connectivity or expensive data transfer fees.

- AI and ML: Utilizing AI and ML algorithms facilitates in automating procedures, extracting useful insights from IoT data, and gradually increasing system intelligence [16].

## 2.4 The emergence of EC

EC represents a distributed computing approach where data processing, storage, and analysis occur in proximity to the data source, rather than exclusively relying on centralized data centers or cloud services [17]. In this paradigm, data processing occurs locally, typically on edge devices or servers positioned near the network edge where data is generated. Figure 2.2 depicts the convergence of EC in IoT by showing the connectivities among edge layers, IoT sensors to connected devices, local porting, mobile devices, cloud servers, edge servers, and local processing.

The key characteristics of EC include the following:

- Proximity to Data Source: Bringing computing resources nearer to data generation points, EC diminishes latency and enhances real-time processing abilities. This closeness is particularly advantageous for applications

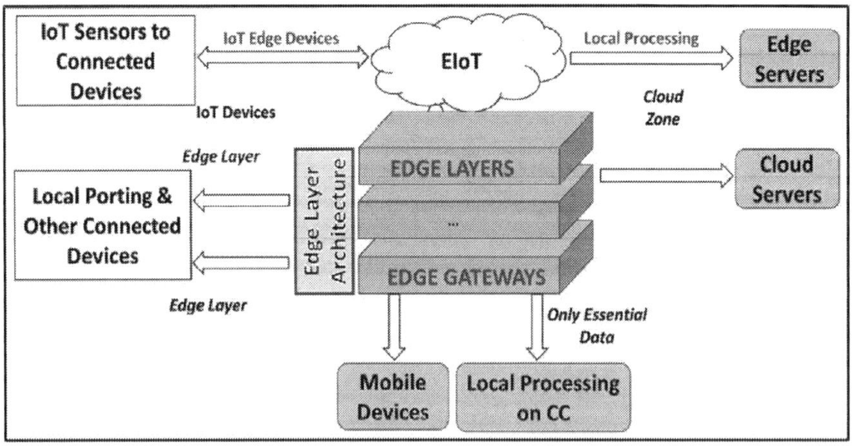


*Figure 2.2* Convergence of EC in IoT.

necessitating swift responses, like industrial automation, autonomous vehicles, and Augmented Reality (AR).

- Decentralization: EC decentralizes computing resources by distributing processing tasks across multiple edge devices or servers. This approach helps alleviate the burden on centralized data centers and reduces reliance on network connectivity, making edge deployments more resilient to network outages or bandwidth limitations.
- Data Localization: EC enables organizations to process and store data locally, which can be advantageous for applications with data sovereignty or compliance requirements. By keeping sensitive data within specific geographic regions or jurisdictions, organizations can ensure compliance with regulatory policies and protect data privacy.
- Scalability and Flexibility: EC architectures are designed to scale horizontally, allowing organizations to easily deploy additional edge nodes or devices as needed to accommodate fluctuating workloads or expanding IoT deployments. This scalability and flexibility enable EC to adapt to evolving business requirements and technological advancements.
- Integration with Cloud Services: While EC enables local data processing and analysis, it often complements rather than replaces centralized cloud services. EC solutions can integrate with cloud platforms to create hybrid architectures that leverage the scalability, storage, and advanced analytics capabilities of the cloud while processing time-sensitive data locally at the edge.
- Security and Privacy: EC presents distinctive security hurdles, especially in dispersed settings featuring numerous edge devices or endpoints. Critical security measures like encryption, access control, authentication, and secure communication protocols are imperative to uphold data integrity and confidentiality within edge implementations. Furthermore, EC frameworks should integrate privacy-enhancing methodologies to adhere to data protection regulations and uphold user privacy.

## 2.5 Benefits of EC

EC provides a compelling array of advantages, empowering organizations to seize fresh opportunities, enhance operational efficiency, and provide groundbreaking services and experiences across various applications and industries [18]. Here are some key advantages:

- Reduced Latency: Compared to conventional cloud-based methods, EC drastically lowers latency by processing data closer to its source. For real-time applications where even milliseconds of latency can affect performance and user experience, such as industrial automation, driverless cars, and AR/Virtual Reality (VR) experiences, this is very advantageous.

- Improved Performance: EC enables faster processing and response times, leading to improved overall system performance. By performing computations locally, edge devices can handle data-intensive tasks more efficiently, leading to faster insights, decision-making, and actions.
- Bandwidth Optimization: By minimizing the need to send massive amounts of data to centralized data centers via the network, EC lowers network congestion and bandwidth consumption. This is particularly helpful in situations like remote or IoT deployments where network connectivity is scarce or unpredictable.
- Enhanced Security: EC mitigates security risks associated with transmitting sensitive data over the network by processing and storing data locally. This approach reduces exposure to potential cyber threats and unauthorized access, enhancing data privacy and confidentiality.
- Scalability and Flexibility: EC architectures are highly scalable and flexible, allowing organizations to deploy additional edge nodes or devices as needed to accommodate growing workloads or evolving business requirements. This scalability enables edge deployments to adapt to changing demand patterns and technological advancements over time.
- Resilience and Reliability: EC architectures are inherently resilient to network outages or disruptions since they can continue to operate autonomously even when disconnected from the centralized cloud infrastructure. This resilience ensures uninterrupted operation and data availability, particularly in mission-critical applications.
- Data Sovereignty and Compliance: Organizations can handle issues with data sovereignty, privacy laws, and compliance requirements by using EC to process and store data locally. By keeping sensitive data within specific geographic regions or jurisdictions, organizations can ensure compliance with regulatory policies and protect data privacy.
- Cost Optimization: EC can help optimize costs by reducing the need for expensive network bandwidth and centralized infrastructure. By distributing computing resources closer to the edge, organizations can minimize data transfer costs and optimize resource utilization, leading to cost savings over time.

## 2.6 The intersection of EC and IoT

The synergy between EC and IoT creates a powerful platform for organizations to capitalize on emerging opportunities, streamline operations, and pioneer novel services across diverse sectors and applications. By harnessing the capabilities of distributed computing at the network edge, organizations can harness the complete potential of IoT to catalyze digital transformation and cultivate value in the interconnected landscape [19]. The convergence of EC and IoT is a transformative force, offering a multitude of benefits that extend beyond traditional boundaries. The following are the ways in which these two technologies intersect and synergize to propel innovation and efficiency in various domains:

- Real-time Data Processing: At the network's periphery, IoT devices produce extensive data volumes, frequently in distant or resource-limited settings. EC empowers organizations to locally analyze these data in real time, facilitating quicker insights, decision-making, and actions without dependence on centralized cloud infrastructure. This capability proves indispensable in applications like industrial automation, smart cities, and connected vehicles, where prompt responses are imperative.
- Reduced Latency: Through processing data near its origin, EC markedly diminishes latency in contrast to conventional cloud-based methods. This feature is particularly advantageous for latency-sensitive IoT applications like industrial control systems, where even minor delays can adversely affect performance and safety.
- Bandwidth Optimization: EC mitigates the necessity of transmitting extensive quantities of IoT data across the network to centralized data centers, thus lowering bandwidth consumption and network congestion. This proves especially advantageous in situations with restricted or inconsistent network connectivity, such as deployments of IoT in remote rural areas or industrial facilities.
- Scalability and Flexibility: EC frameworks exhibit remarkable scalability and flexibility, allowing organizations to introduce supplementary edge nodes or devices as required to accommodate expanding IoT deployments. This scalability empowers IoT ecosystems to adjust to shifting demand dynamics and expand efficiently, all while avoiding undue strain on centralized infrastructure.
- Edge AI and ML: EC facilitates the implementation of AI and ML models directly on IoT devices or edge servers, facilitating intelligent data processing and analytics at the edge. This empowers functionalities like predictive maintenance, anomaly detection, and real-time optimization without dependence on cloud-based AI services, thereby bolstering privacy and mitigating latency.
- Local Decision-making: EC empowers IoT devices to autonomously make local decisions based on real-time data analysis, eliminating the necessity for continuous communication with centralized cloud services. This capability ensures that IoT applications can maintain operations even during connectivity disruptions, thereby enhancing resilience and reliability.
- Security and Privacy: By reducing exposure to potential cyber threats and unauthorized access, EC strengthens security and privacy in IoT installations. Organizations may reduce the risk of data breaches and maintain regulatory compliance by processing and storing data locally, protecting user privacy and sensitive IoT data.

## 2.7 Complementary strengths of IoT and EC

IoT and EC are complementary technologies that offer unique strengths and capabilities, which, when combined, enable organizations to build more powerful and efficient systems [20]. The following are the complementary strengths of IoT and EC.

a. Strengths of IoT

- Data Generation: Using sensors and actuators, IoT devices are excellent at gathering enormous volumes of data from the real world [21]. These gadgets may track a number of variables, including motion, temperature, humidity, pressure, and more, giving important information about the surroundings and operations.
- Connectivity: Wi-Fi, Bluetooth, Zigbee, RFID, cellular networks, LoRaWAN, and other communication protocols are used by IoT devices to connect to the internet or local networks. IoT devices can send data to other edge devices or centralized systems for additional processing and analysis thanks to this connectivity.
- Scalability: Because of the great scalability of IoT systems, businesses may deploy and oversee thousands or even millions of devices in a variety of settings. IoT implementations can handle increasing data volumes and support growing use cases and applications thanks to this scalability.
- Diverse Applications: IoT technology is used in a wide range of sectors and fields, such as healthcare, agriculture, transportation, smart cities, smart homes, and industrial automation. From consumer electronics to vital industrial applications, IoT solutions can handle a broad variety of use cases.
- Remote Monitoring and Control: IoT devices enable remote monitoring and control of assets, equipment, and processes, allowing organizations to optimize operations, improve efficiency, and reduce maintenance costs. Remote monitoring capabilities are particularly valuable in scenarios where physical access is limited or impractical.

b. Strong Point of EC

- Low Latency: By processing data close to its source, EC reduces latency and makes it possible to react to events and interactions in real time or almost real time. For latency-sensitive applications like driverless cars, industrial control systems, and AR, this feature is essential.
- Reduced Bandwidth Usage: EC minimizes the need to transmit large volumes of data over the network to centralized data centers, reducing bandwidth usage and network congestion. This is especially beneficial in scenarios with limited or expensive network connectivity.
- Resilience and Reliability: EC architectures are resilient to network outages or disruptions since they can continue to operate autonomously

even when disconnected from centralized cloud infrastructure. This resilience ensures uninterrupted operation and data availability, particularly in mission-critical applications.
- Privacy and Compliance: EC enhances privacy and compliance by processing and storing data locally, reducing exposure to potential cyber threats and unauthorized access. This approach helps organizations comply with regulatory requirements and protect sensitive data and user privacy.
- Local Decision-making: EC enables devices to independently make local decisions through real-time data analysis, eliminating the requirement for continual communication with centralized cloud services. This facilitates quicker responses to events and interactions, thereby enhancing system performance and efficiency

## 2.8 Role of EC in IoT applications

EC complements traditional cloud-based IoT architectures by bringing computation and storage capabilities closer to the edge of the network, thereby improving performance, reliability, and security while enabling new applications and use cases, as shown in Table 2.1. EC plays a crucial role in enhancing the capabilities of IoT systems in several ways:

- Reduced Latency: EC shortens the distance data must travel between its generation point and processing location, resulting in reduced latency. This is particularly crucial for applications necessitating instantaneous responses, like industrial automation or autonomous vehicles.
- Bandwidth Optimization: EC aids in reducing the volume of data that necessitates transmission to the cloud for processing and storage. Through filtering and aggregating data at the edge, solely pertinent information is conveyed, thereby diminishing bandwidth usage and related expenses.
- Improved Reliability: EC can provide localized processing and storage capabilities even when internet connectivity is limited or unreliable. This ensures that IoT systems remain functional and responsive even in environments with intermittent network coverage.
- Enhanced Privacy and Security: Conducting sensitive data processing at the edge offers a solution to privacy apprehensions by maintaining data within a controlled environment, rather than transmitting it to distant cloud servers [22]. Furthermore, EC facilitates real-time data encryption and analysis, bolstering security measures.
- Scalability: EC disperses computational responsibilities among a network of edge devices, enabling IoT systems to scale with greater efficiency. This adaptability proves especially advantageous for applications characterized by dynamic workloads or evolving requirements.
- Autonomy: EC enables IoT devices to make local decisions and take immediate actions without relying on constant communication with a central

*Table 2.1* Key Differences between EC and IoT

| *SN* | *EC* | *IoT* |
|---|---|---|
| **1** | An approach to computing that does not rely on a centralized data-processing site, such as the cloud, but instead processes data at or close to the data source. | Smart operations are made possible by a network of interconnected devices that gather, share, and process data online. |
| **2** | Minimizes network strain and decreases latency by processing data closer to the data source. | Collects and transmits data from devices for monitoring, automation, and analytics. |
| **3** | Primarily uses edge devices to process data locally before forwarding the required info to centralized systems. | Gathers information from sensors and devices and usually transmits it to edge devices or the cloud for processing. |
| **4** | On-site processing makes it quick with low latency, which is essential for real-time applications. | Higher delays could occur, contingent on cloud connectivity and network speed. |
| **5** | Can function with sporadic or restricted connectivity because data processing takes place locally. | Often requires a stable internet connection for real-time data collection and remote access. |
| **6** | Enhanced security by keeping sensitive data on local devices, reducing the exposure to cloud-based risks. | Requires robust security measures for data transmission to and from cloud platforms, which may be susceptible to attacks. |
| **7** | Reduces the amount of data sent to the cloud by processing it locally and filtering out unnecessary data. | Typically sends a large volume of data to the cloud, often requiring significant storage and processing capabilities. |
| **8** | May require more power at the edge due to local processing demands. | IoT devices are generally optimized for low power consumption, often relying on the cloud for processing. |

server. This enhances autonomy and resilience, especially in scenarios where network connectivity is intermittent or unavailable.

- Customization and Personalization: EC facilitates the customization of IoT applications based on local conditions and requirements. By processing data closer to the source, edge devices can adapt quickly to changing environments and user preferences, providing more personalized experiences.

## 2.9 Challenges in integrating EC and IoT

Integrating EC with IoT presents several challenges, including the following:

- Heterogeneity of Edge Devices: Edge devices come in various forms with different computational capabilities, communication protocols, and operating systems. Integrating these heterogeneous devices into a cohesive EC ecosystem requires standardization efforts and interoperability solutions.

- Data Management and Processing: Edge devices generate large volumes of data, often in real time. Efficiently managing and processing these data at the edge while ensuring low latency and high reliability is challenging. It requires optimizing algorithms, deploying edge analytics, and balancing computational loads across the edge network.
- Security and Privacy: Edge devices are often deployed in physically insecure environments, making them vulnerable to security threats such as physical tampering, unauthorized access, and data breaches. Implementing robust security measures, including encryption, access control, and secure bootstrapping, is crucial to protect sensitive data and ensure the integrity of EC systems [23].
- Network Connectivity and Reliability: Edge devices may operate in environments with limited or intermittent network connectivity, such as remote locations or industrial settings. Ensuring reliable communication between edge devices and centralized systems, as well as addressing network latency and bandwidth constraints, requires deploying resilient networking solutions and edge caching mechanisms [24].
- Scalability and Management: Effectively managing a considerable quantity of distributed edge devices on a grand scale presents multifaceted challenges encompassing configuration, monitoring, and software updates. As the number of these devices proliferates, the complexity of overseeing them concurrently amplifies. To address this, organizations often implement centralized management platforms and leverage automation tools to streamline these operations. These platforms enable administrators to configure settings, monitor device health, and deploy software updates across the distributed network efficiently.
- Integration with Cloud Services: While EC offloads computational tasks from centralized cloud servers, seamless integration between edge and cloud environments is essential for data synchronization, orchestration, and hybrid processing. Bridging the gap between edge and cloud infrastructures requires standardized interfaces, data synchronization protocols, and efficient resource allocation mechanisms.
- Regulatory and Compliance Requirements: EC deployments must comply with regulatory frameworks and industry standards governing data privacy, security, and interoperability. Navigating these legal and compliance requirements, especially in highly regulated sectors such as healthcare and finance, adds complexity to EC integration efforts.

Addressing these challenges requires collaboration between stakeholders across the EC ecosystem, including device manufacturers, software developers, network operators, and regulatory bodies. By overcoming these obstacles, organizations can unlock the full potential of EC and realize the benefits of distributed intelligent IoT systems.

## 2.10 Challenges and future directions

The integration of EC with IoT faces several challenges and is poised for exciting future developments. By addressing challenges and embracing future directions, EC and IoT are poised to revolutionize industries, enhance user experiences, and enable a new era of connected intelligence.

### *2.10.1 Challenges*

- Standardization: The lack of standardized protocols and interfaces for EC and IoT devices hinders interoperability and seamless integration. Efforts to develop industry-wide standards are essential for addressing this challenge.
- Security and Privacy: Safeguarding edge devices and data against security threats while ensuring compliance with privacy regulations remains an enduring challenge. To address this, comprehensive security measures, such as encryption, authentication, and access control, must be meticulously integrated throughout EC ecosystems. These measures serve as critical safeguards, fortifying the integrity and confidentiality of data traversing edge networks. By diligently implementing these security protocols, organizations can fortify their edge environments against potential vulnerabilities and uphold the privacy rights of users and stakeholders.
- Scalability: Effectively handling the increasing volume of edge devices at scale demands scalable solutions for device provisioning, configuration management, and software updates. Automation tools and centralized management platforms emerge as indispensable resources in streamlining these operations. These solutions facilitate the efficient provisioning of devices, ensuring they are configured appropriately and equipped with the necessary software updates. By leveraging automation and centralized management, organizations can effectively manage their expanding fleet of edge devices, optimizing performance and enhancing operational efficiency.
- Network Connectivity: Edge devices often operate in environments characterized by limited or intermittent network connectivity, presenting challenges for data transmission and synchronization. To address these hurdles, deploying resilient networking solutions and implementing edge caching mechanisms prove essential. Resilient networking solutions ensure robust connectivity by leveraging technologies such as mesh networking or satellite communication to maintain communication even in adverse conditions. Additionally, edge caching mechanisms store frequently accessed data locally on the device, reducing the dependency on continuous network connectivity and enhancing data availability. By integrating these strategies, organizations can mitigate the challenges posed by unreliable network environments and ensure seamless operation of edge devices.

- Data Management: Edge devices generate vast amounts of data, requiring efficient storage, processing, and analysis capabilities. Edge analytics and data filtering techniques are essential for extracting actionable insights from raw sensor data.

### 2.10.2 *Future directions*

- Edge AI: Integrating AI and ML algorithms at the edge facilitates real-time decision-making and predictive analytics. This approach, known as Edge AI, empowers IoT devices to analyze and process data locally, thereby reducing latency and minimizing reliance on centralized cloud servers. By harnessing the computational power of edge devices, organizations can enhance responsiveness and efficiency in their IoT applications while preserving data privacy and minimizing bandwidth usage.
- Fog Computing: Fog computing extends the capabilities of EC by utilizing intermediary nodes positioned between edge devices and centralized cloud servers. This strategy facilitates hierarchical data processing and distributed intelligence, thereby bolstering scalability and flexibility within the network architecture. By introducing intermediary nodes into the computing infrastructure, organizations can optimize resource utilization and enhance the responsiveness of their IoT applications while maintaining a decentralized and adaptable framework.
- 5G and EC: The rollout of 5G networks promises to revolutionize EC by delivering ultra-low latency and high bandwidth connectivity. EC combined with 5G enables applications such as autonomous vehicles, AR, and remote healthcare with unprecedented speed and reliability.
- Edge-to-Cloud Orchestration: Seamless orchestration and workload migration between edge and cloud environments optimize resource utilization and application performance. Edge-to-cloud orchestration platforms enable dynamic workload management and hybrid processing, balancing computation between edge devices and centralized cloud servers.
- Edge-as-a-Service (EaaS): EaaS models offer on-demand provisioning of EC resources and services, enabling enterprises to deploy and manage edge applications cost-effectively. EaaS platforms abstract the complexities of edge infrastructure, allowing organizations to focus on developing innovative IoT solutions.
- Distributed Ledger Technologies: Integrating distributed ledger technologies (DLT), such as blockchain, with EC enhances data integrity, transparency, and auditability in IoT applications. DLT ensures tamper-proof storage and secure sharing of IoT data across distributed edge nodes, fostering trust and accountability in decentralized ecosystems.
- EC in Industry Verticals: Tailoring EC solutions to specific industry verticals, such as manufacturing, healthcare, and transportation, unlocks new opportunities for innovation and efficiency. Industry-specific EC platforms

and use cases address unique requirements and challenges, driving digital transformation and competitive advantage.

## 2.11 Conclusions

The fusion of EC and IoT heralds a transformative era in digital technology, promising unprecedented innovation and efficiency across various sectors. EC's proximity to data sources addresses many constraints of traditional cloud-based models, while IoT interconnects devices, enabling intelligent, automated operations. This synergy not only enhances system capabilities but also tackles challenges inherent in centralized computing models. Understanding the interplay between EC and IoT is paramount for developing and deploying next-generation applications. This chapter has delved into their interdependent relationship, exploring how integration leads to more efficient, responsive, and secure systems.

However, this integration is not without hurdles. Challenges such as standardization, security, scalability, and regulatory compliance must be addressed. Collaboration among stakeholders is crucial to overcome these obstacles and realize the full potential of distributed intelligent IoT systems. Looking ahead, the future holds promise with advancements like edge AI, fog computing, 5G integration, and EaaS models. These developments will further optimize performance, scalability, and security, unlocking new opportunities for innovation and efficiency. In conclusion, the convergence of EC and IoT marks a paradigm shift, ushering in an era of unprecedented connectivity and intelligence. By addressing challenges and embracing future directions, organizations can harness the transformative power of EC and IoT to revolutionize industries and enhance user experiences.

**Acknowledgements:** The authors would like to acknowledge their colleagues and friends who supported and encouraged them to carry out this research work.

## References

1. Rasheed, Z., Ma, Y.-K., Ullah, I., et al. 2024. Edge computing in the digital era: The Nexus of 5G, IoT and a seamless digital future. In *Future Communication Systems Using Artificial Intelligence, Internet of Things and Data Science*, 1–22. CRC Press.
2. Mahajan, M. P., Nikam, R. R., Patil, V. P., et al. 2017. Smart refrigerator using IOT. *International Journal of Latest Engineering Research and Applications* 2 (March): 86–91. https://d1wqtxts1xzle7.cloudfront.net/99543321/31.201703127-libre.pdf
3. Kranenburg, R. V. 2007. The internet of things. In *A Critique of Ambient Technology and the All-Seeing Network of RFID*. Network Notebooks 2. Institute of Network Cultures, Amsterdam.

4. Choudhary, V., and Taruna, S. 2020. The highly secure polynomial pool-based key pre-distribution scheme for wireless sensor network. *Journal of Discrete Mathematical Sciences and Cryptography* 23(1): 95–114. https://doi.org/10.1080/09720529.2020.1721880
5. Yang, Y., Luo, X., Chu, X., et al. 2020. *Fog-Enabled Intelligent IoT Systems*. Springer Nature, Cham.
6. Anorboyevich, A. R. 2024. *Exploring the Evolutionary Path: A Historical Analysis of Internet of Things (IoT) Technologies*. Global Book Publishing Services.
7. Munirathinam, S. 2020. Industry 4.0: Industrial Internet of Things (IIOT). In *Advances in Computers*, ed. P. Raj and P. Evangeline, 129–164. Elsevier.
8. Botterman, M., Cave, J., and Doria, A. 2020. Standardization Issues. In *ICT Policy, Research, and Innovation: Perspectives and Prospects for EU-US Collaboration*, ed. S. Klessova, S., Engell, S., M. Botterman, and J. Cave, 309–330. Wiley.
9. Deng, S., Zhao, H., Fang, W., et al. 2020. Edge intelligence: The confluence of edge computing and artificial intelligence. *IEEE Internet of Things Journal* 7, no. 8 (August): 7457–7469.
10. Chang, Z., Liu, S., Xiong, X., et al. 2021. A survey of recent advances in edge-computing-powered artificial intelligence of things. *IEEE Internet of Things Journal* 8(18, September): 13849–13875.
11. Allioui, H., and Mourdi, Y. 2023. Exploring the full potentials of IoT for better financial growth and stability: A comprehensive survey. *Sensors* 23(19): 1–68.
12. Verma, S., Kawamoto, Y., Fadlullah, Z. M., et al. 2017. A survey on network methodologies for real-time analytics of massive IoT data and open research issues. *IEEE Communications Surveys & Tutorials* 19(3): 1457–1477.
13. Grammatikis, P. I. R., Sarigiannidis, P. G., and Moscholios, I. D. 2019. Securing the Internet of Things: Challenges, threats and solutions. *Internet of Things* 5, 41–70.
14. Noura, M., Atiquzzaman, M., and Gaedke, M. 2019. Interoperability in Internet of Things: Taxonomies and open challenges. *Mobile Networks and Applications* 24: 796–809.
15. Rajasegar, R. S., Gouthaman, P., Ponnusamy, V., et al. 2024. Data privacy and ethics in data analytics. In *Data Analytics and Machine Learning, Studies in Big Data*, ed. P. Singh, A. R. Mishra, and P. Garg, vol. 145, 195–213. Springer, Singapore.
16. Campbell, C., Sands, S., Ferraro, C., et al. 2020. From data to action: How marketers can leverage AI. *Business Horizons* 63(2): 227–243.
17. Yu, W., Liang, F., He, X., et al. 2018. A survey on the edge computing for the Internet of Things. *IEEE Access* 6: 6900–6919.
18. Raj, P., Akilandeswari, J., and Marimuthu, M. 2021. The edge AI paradigm: Technologies, platforms and use cases. *Advances in Computers* 127: 139–182.
19. Pan, J., and McElhannon, J. 2018. Future edge cloud and edge computing for Internet of Things applications. *IEEE Internet of Things Journal* 5(1, February): 439–449.
20. Qiu, T., Chi, J., Zhou, X, et al. 2020. Edge computing in industrial Internet of Things: Architecture, advances and challenges. *IEEE Communications Surveys & Tutorials* 22(4): 2462–2488.
21. Choudhary, V., and Taruna, S. 2019. A distributed key management protocol for wireless sensor network. In *Advanced Informatics for Computing Research, Communications in Computer and Information Science*, ed. A. Luhach, D. Singh, P. A. Hsiung, K. Hawari, P. Lingras, and P. Singh, 243–256. Springer, Singapore.

22. Choudhary, V., Taruna, D. S., and Purbey, L. B. 2018. A comparative analysis of cryptographic keys and security. In *3rd International Conference and Workshops on Recent Advances and Innovations in Engineering*, 1–8. IEEE.
23. Choudhary, V., and Taruna, S. 2021. An intrusion detection technique using frequency analysis for wireless sensor network. In *International Conference on Computing, Communication, and Intelligent Systems*, 206–210. IEEE.
24. Gupta, S., Bharti, P. K., and Choudhary, V. 2011. Fuzzy logic based routing algorithm for mobile ad hoc networks. In *High Performance Architecture and Grid Computing, Communications in Computer and Information Science*, ed. A. Mantri, S. Nandi, G. Kumar, and S. Kumar, 574–579. Springer, Berlin, Heidelberg.
25. Puri, S., Babu, M. A., Ahmmed, M. M., et al. 2024. An efficient fuzzy colored Petri-nets-based ubiquitous framework for diversified culture of building automation in India. *Procedia Computer Science* 236: 476–484. https://www.sciencedirect.com/science/article/pii/S187705092401072X
26. Kaushik, S., and Puri, S. 2012. Online transaction processing using enhanced sensitive data transfer security model. In *Students' Conference on Engineering and Systems*, 1–5. IEEE Press.
27. Puri, S., and Singh, S. P. 2020. A fuzzy matching based image classification system for printed and handwritten text documents. *Journal of Information Technology Research* 13(2): 155–194. https://doi.org/10.4018/JITR.2020040110

Chapter 3

# Foundations of edge intelligence

*Avnish Singh and Govind P. Gupta*

## 3.1 Introduction to edge intelligence

The Internet of Things (IoT) devices, smart sensors, and data-intensive applications have changed the generation, processing, and usage of data with intensities based on the number of industries relying on a large number of streams of big data in real time. This includes self-driving vehicles and smart city systems. Traditional cloud computing, whereby data are primarily processed using centralized servers, cannot capture some of the desirable qualities in terms of speed, privacy, and efficiency required for the modern innovations witnessed today. However, the traditional cloud models suffer from problems such as high latency and low bandwidth, issues of privacy, and dependency on an uninterrupted connection to the cloud [1].

In the face of these challenges, edge intelligence (EI) offers the possibility of embedding artificial intelligence (AI) capabilities directly into devices at the point where data are created. Its essence entails reducing latency and dependency on network connections through the processing of information locally within individual devices. The applications requiring the least delay and the utmost privacy are, in fact, those that EI has special benefits for: industrial automation, health diagnostics, and autonomous driving.

The most significant advantage of EI is that it reduces the latency by quite a lot, because data processing takes place at the source. For example, in applications such as autonomous driving, decisions that could determine a potential collision are split seconds away. Therefore, computation ought to be immediate. Privacy and security are also improved because sensitive data stay local, which is in line with new regulations in healthcare and finance where it is focused on eliminating the sending of sensitive information over networks.

EI also saves resources by minimizing the overhead at the cloud servers and reducing bandwidth consumption as processes are carried out at the edge level. The distributed system comprises device, edge, network, and cloud layers, thus enhancing efficiency, resiliency, and economics. If the network breaks, the devices remain operational without further disrupting the continuity of the critical services.

DOI: 10.1201/9781003512066-3

In summary, EI addresses some of the limitations of the traditional cloud computing models and expands technological capability by moving AI closer to the point of origin of data. This is revolutionizing many fields, allowing one to make decisions faster in an autonomous vehicle, for instance, as well as allowing for real-time diagnostics in the health devices that people wear around their bodies.

## 3.2 Architecture of edge intelligence-based systems

The EI architecture is composed of multiple layers, each of which represents a need in the pipeline for data processing and decision-making [2]. In this manner, each layer's unique characteristics help to build a cohesive and high-performing ecosystem taken as a whole.

The layers of the EI-based system are as follows:

a) **Device Layer:** The lowest layer is the Device Layer, which captures the data generation and acts as a collection point. It contains IoT devices for industrial and consumer applications, sensors for environmental and process monitoring, mobile devices for user interaction and data collection, and connected vehicles for transportation-related data. These edge devices will collect raw information from the physical world and forward it to the other layers of the system for processing.
b) **Edge Layer:** This is the first layer where initial data processing and analysis occur. The Edge Layer includes real-time processing for immediate data handling, temporary storage for caching, quick analytics for rapid insights, and local decisions for autonomous operations. This layer reduces latency while offloading the central cloud infrastructure by bringing insights and making time critical decisions as close to where the data originates as possible.
c) **Network Layer:** It acts as a middleman between the Edge Layer and the Cloud Layer where it allows data to be transferred efficiently and securely. It relies on load balancers for distribution of traffic, traffic management to optimize routing, security controls in order to protect it, and content delivery networks to deliver content effectively. It ensures that the data transfer occurs efficiently and effectively while managing the traffic and improving the performance of the system.
d) **Cloud Layer**: At the top of the architecture, the Cloud Layer provides central processing and storage. It performs long-term storage for data archiving, big data analytics for complex data processes, artificial intelligence/machine learning (AI/ML) for training complex models and model deployment, and resource orchestration to manage cloud resources. The cloud's nearly unlimited computing power and storage capacity allow the system to reveal deeper insights and create more sophisticated AI-driven capabilities.

Thus, the interplay and cooperation of these four layers are considered the real power of the edge intelligence architecture, which offers unique new capabilities for each that synergize into an impressively efficient and responsive system, as depicted in Figure 3.1.

The Device Layer is the bottom layer, which produces the raw data that feeds the entire system. The Edge Layer assumes this critical role of real-time processing and decision-making, extracting insights, and triggering

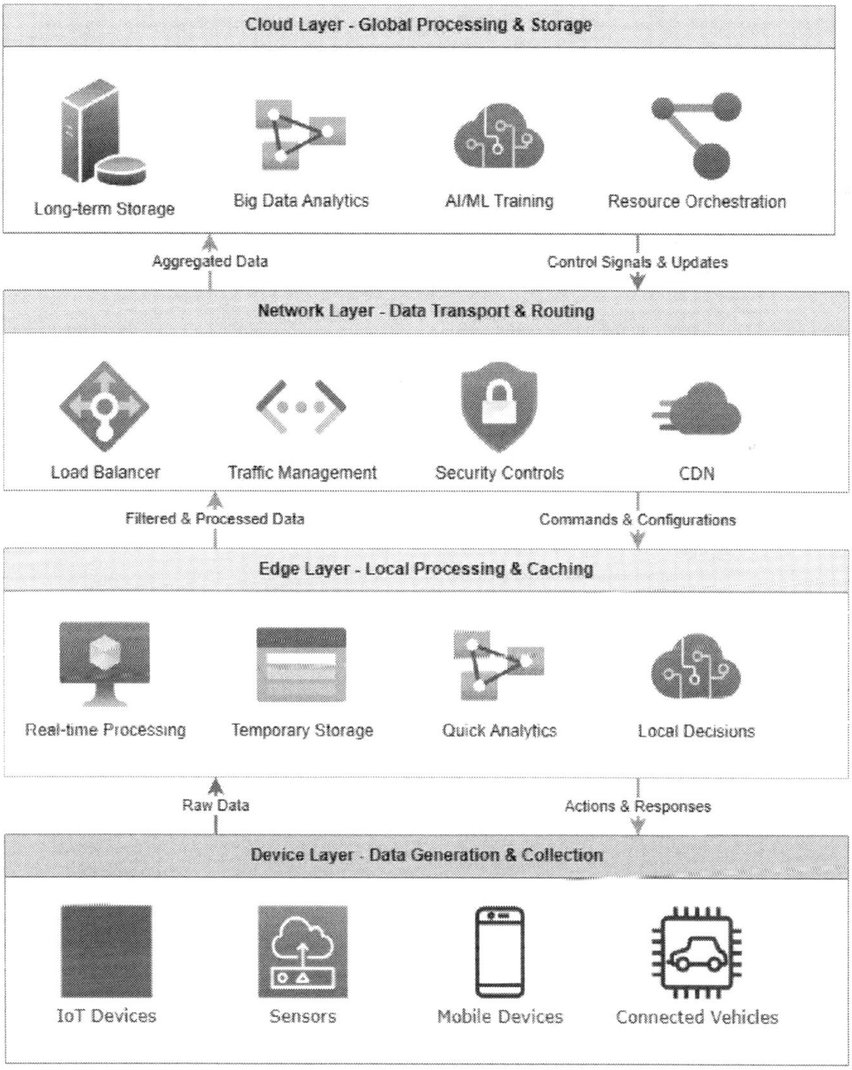


*Figure 3.1* Architecture of edge intelligence–based system.

immediate actions without the latency of round-trip communication to the cloud. Through optimization and maximizing data flow, the Network Layer is responsible for ensuring safe and reliable data transmission between the edge and the cloud. This layer exploits load balancing, traffic management, and content delivery networks, thus making the whole system more performance-oriented and responsive. Finally, on top of the architecture, the Cloud Layer aggregates data to carry out advanced analytics and to train sophisticated ML models, while orchestrating the allocation of computing resources. This hub of intelligence and processing power complements the real-time capabilities of the Edge Layer, unlocking the full potential of EI.

## 3.3 Benefits of edge intelligence

Because of the distributed and layered nature of this EI architecture, it brings into use a myriad of benefits accounting for the limitations of traditional models being cloud-centric, including the following:

1. **Reduced Latency:** Processing data nearer to the source and then taking decisions means the latency stands significantly reduced in the Edge Layer, bringing in real-time responsiveness to applications that require prompt responses at times, including in industrial automation or autonomous vehicles [3].
2. **Enhanced privacy and security:** The decentralized processing and storage of information at the edge minimize the movement of sensitive data to the cloud, thus enhancing privacy and security about the same. This answers many data sovereignty issues and other concerns with regulatory compliance [3].
3. **Optimized Usage of Resources**: This means offloading of relevant processing tasks to the Edge Layer, so the utilization of the Cloud Layer can be optimized to cut costs and enhance the general performance of the system, especially considering the big requirements of the application in terms of processing [4].
4. **Improved Resilience and Fault Tolerance:** As it is an architecture of distributed nature, the system has intrinsic resilience. So, the system would continue to work in the case of a partial network failure or a cloud outage, thus ensuring service delivery without a hitch.
5. **Enhanced Scalability:** The architecture of EI can be modular and supports seamless scaling on both frontiers: new edge devices and scalability to enlarge the processing capability of the cloud layer. Organizations can adapt to changing data requirements and growth [3].

This architecture, in combining those layers and their functions seamlessly, equips organizations to make the most of the power of data at the edge: to deliver innovative, responsive, and efficient services tailored for today's digital landscape.

## 3.4 Software framework for edge intelligence

There are now specialized software frameworks, each optimized for the resource-constrained environments that are typical in the end applications in which much of the edge AI will run—holding limited computational power, memory, and connectivity. They are designed to make models easier to build and deploy on edge devices supporting lightweight design, interoperability, and modularity.

- **TensorFlow Lite:** TensorFlow Lite [5] is TensorFlow tailored for mobile and embedded environments. With it, one can run inference on very low-power devices like smartphones, sensors, and modules in the IoT environments. This framework supports model quantization, reducing the precision of model parameters to fit the constraint of the edge device, thereby massively reducing the size of the model and computations without significant loss in accuracy. Applications powered in real time on edge devices from this tool comprise image recognition, speech processing, natural language understanding, and so forth.
- **EdgeX Foundry:** This is an open-source middleware platform designed to provide interoperability between different IoT devices and various forms of data processing services. The modular architecture allows EdgeX Foundry [6] to be very adaptable to heterogeneous edge environments, integrating with a wide range of hardware and software. EdgeX Foundry supports applications requiring fast, local decision-making such as real-time environmental monitoring and industrial automation. It can collect, aggregate, and manage real-time data.
- **Amazon Web Services (AWS) IoT Greengrass:** AWS IoT Greengrass [7] extends cloud capabilities to edge devices enabling local data processing, messaging, and ML tasks even when the device is disconnected from the network. With support for over-the-air (OTA) updates, Greengrass makes it easy to deploy and manage models in remote or geographically dispersed environments. This makes Greengrass a pretty good fit for industrial IoT scenarios where such frequent model updates and version control are essential.

Each of these frameworks consumes resources efficiently and is scalable. Building blocks, that support modular components, help developers to adapt solutions based on the needs of an application, which makes it possible to have highly diverse, distributed edge networks, and creates a very robust system for EI applications.

## 3.5 Techniques to reduce complexity of models for edge intelligence

All AI models deployed at the edge need to be tuned to the constraints under which edge devices function – limited memory, computational resources and power. Popular models are deep neural network (DNN) [3, 8], convolutional neural network (CNN) [9] for image recognition or object detection, and recurrent neural network (RNN) [10] for sequential data processing. Optimization of these models is the key towards realizing efficient EI without losing performance.

1. **Quantization:** Instead of reducing model size, it expresses reduction in terms of a decrease in the precision of model parameters. Parameters are converted from 32 bit-float numbers into 8-bit integers by quantization [11] to reduce memory requirements and speed up computation with minimal loss of accuracy. Its uses are very common in models running on smartphones, IoT devises, and other hardware that have limited computing power.
2. **Pruning:** Pruning [12] eliminates redundant weights or neurons in a model, which reduces the total size of the model and improves energy efficiency. It has been particularly effective in applications such as wearable devices and industrial IoT sensors where power consumption is a limiting factor, and the end device needs to have a large battery life.
3. **Knowledge distillation:** This trains a more compact, efficient student model to mimic the behaviour of a more resource-intensive teacher model [13]. The student model can now perform tasks with near-equivalent accuracy but with far fewer resources, making it suitable for edge deployments with limited processing capabilities.
4. **Federated Learning:** The updating of models, based on decentralized data from numerous edge devices, is done without transferring raw data to a central server. Thus, federated learning [14] supports privacy-preserving AI by keeping data localized while sharing only aggregated model updates. Especially for data-sensitive sectors like healthcare and finance, this principle is particularly valuable, because user privacy is fundamental.
5. **Edge-Specific Neural Network Architectures:** Models such as MobileNet [15], SqueezeNet [16], and EfficientNet [17] are specifically designed for the constrained memory of the environment, and thus these models tend to have smaller architectures that tend to be more efficient. They particularly work well for very deep feedforward neural networks in applications like image classification and object detection at the level of mobile and embedded devices.

These techniques are able to deliver high-end AI functionality on constrained devices, thereby allowing for almost instant decisions and significantly reduced reliance on the cloud. It will be possible to attain optimal performance in EI through knowledge distillation, federated learning, pruning, and quantization in applications.

## 3.6 Applications of edge intelligence

EI transforms a cross section of sectors as the edge implementation enables speedy, localized insights and decision-making at the edge device rather than on a cloud resource [18]. This localized processing reduces latency and preserves privacy, making EI invaluable in numerous fields:

1. **Autonomous driving:** The EI processing is critical in processing vast sensor data of an autonomous vehicle, from Radio Detection and Ranging (Radar), Light Detection and Ranging (Lidar), and camera inputs. Developing frameworks that support local object detection and data fusion allow the vehicle to process things within milliseconds of changing road conditions. For example, at such a time, cloud connectivity may not be accessible or even be throttled during real-time navigation and obstacle avoidance.
2. **Healthcare:** The embedded intelligence in wearables and remote health monitoring systems enables real-time diagnostic insights such that sensitive data will stay on the device, thus protecting patient privacy. This is very useful in real-time, unobtrusive monitoring of patients where immediate intervention might be crucial, for example, in monitoring heart rates, blood glucose levels, falls by elderly patients.
3. **Smart Cities:** EI provides infrastructure for smart cities with applications like traffic monitoring, waste management, and public safety. With local processing, EI provides the ability to respond in real time to the dynamic urban challenges, optimize resource use, and improve the efficiency of municipal service delivery – for example, to identify on-the-fly patterns of traffic congestion or manage energy distribution to cut operational costs and improve conditions of living in urban spaces.
4. **Industrial IoT:** EI can be used for predictive maintenance by analysing equipment data on-site in industrial applications. Since mechanical problems are detected early, the chances of losing time and incurring expenses from maintenance are reduced. This is very important for companies in industrial manufacturing, energy production, and logistics since these sectors normally incur costs and productivity loss due to unplanned stops.

These applications indicate the power of EI in developing responsive and data-driven systems across industries where immediacy, privacy, and minimal dependency on cloud resources are needed.

## 3.7 Conclusion

EI is a novel solution to the limitations of traditional cloud computing, which is decentralized in terms of AI processing and decision-making. Bringing computational intelligence closer to data sources will enable real-time insights and actions across a range of applications. A multi-layered edge intelligence architecture comprises the Device, Edge, Network, and Cloud Layers, allowing for a balanced approach that reduces latency, enhances privacy, and optimizes cloud resources. This approach supports high-performance applications across all sorts of areas, from autonomous vehicles to smart healthcare devices where immediacy and security are paramount.

The specific software frameworks and edge-optimized AI models discussed – the approach taken here, for example, such as TensorFlow Lite or federated learning – enable efficient use of computational resources and grant advanced AI capabilities even within the context of resource-limited environments. EI's specific advantages, specifically scalability, resilience, and effective use of resources, underscore this rapidly growing necessity within this well-connected world in which insights must be actionable quickly from data. In reality, though, the broader evolution of EI together with cloud resources opens up an entire horizon of hybrid systems that will allow for much more responsive and intelligent boundaries in any applications, as well as far better systems for industries and users alike.

## References


[1] Shi, Weisong, et al. "Edge computing: Vision and challenges." *IEEE Internet of Things Journal* 3.5 (2016): 637–646.

[2] Xu, Dianlei, et al. "Edge intelligence: Architectures, challenges, and applications." arXiv preprint arXiv:2003.12172 (2020).

[3] Chen, Jiasi, and Xukan Ran. "Deep learning with edge computing: A review." *Proceedings of the IEEE* 107.8 (2019).

[4] Mendez, Javier, et al. "Edge intelligence: Concepts, architectures, applications, and future directions." *ACM Transactions on Embedded Computing Systems (TECS)* 21.5 (2022).

[5] TensorFlow lite: An overview. Access from https://www.tensorflow.org/lite.

[6] EdgeX foundry: A vendor-neutral open source framework for IoT edge computing. Access from https://www.edgexfoundry.org/.

[7] AWS IoT Greengrass developer guide. Access from https://docs.aws.amazon.com/greengrass/v2/developerguide/what-is-iot-greengrass.html.

[8] Data Versity. *A brief history of deep learning*. (2022). Access from https://www.dataversity.net/brief-history-deep-learning/.

[9] Avishek Biswas. *The history of convolutional neural networks for image classification: 1989–today*. Towards Data Science. (2024). Access from https://towardsdatascience.com/the-history-of-convolutional-neural-networks-for-image-classification-1989-today-5ea8a5c5fe20.

[10] *A brief history and introduction of recurrent neural networks*. Towards Data Science (2021). Access from https://towardsdatascience.com/a-battle-against-amnesia-a-brief-history-and-introduction-of-recurrent-neural-networks-50496aae6740.

[11] Jacob, Benoit, et al. "Quantization and training of neural networks for efficient integer-arithmetic-only inference." *Proceedings of the IEEE Conference on Computer Vision and Pattern Recognition* (2018).
[12] Han, Song, Huizi Mao, and William J. Dally. "Deep compression: Compressing deep neural networks with pruning, trained quantization and Huffman coding." arXiv preprint arXiv:1510.00149 (2015).
[13] Hinton, Geoffrey. "Distilling the knowledge in a neural network." arXiv preprint arXiv:1503.02531 (2015).
[14] Li, Tian, et al. "Federated learning: Challenges, methods, and future directions." *IEEE Signal Processing Magazine* 37.3 (2020).
[15] Howard, Andrew G. "Mobilenets: Efficient convolutional neural networks for mobile vision applications." arXiv preprint arXiv:1704.04861 (2017).
[16] Iandola, Forrest N. "SqueezeNet: AlexNet-level accuracy with 50x fewer parameters and< 0.5 MB model size." arXiv preprint arXiv:1602.07360 (2016).
[17] Tan, Mingxing, and Quoc Le. "Efficientnet: Rethinking model scaling for convolutional neural networks." *International Conference on Machine Learning*. PMLR (2019).
[18] Zhou, Zhi, et al. "Edge intelligence: Paving the last mile of artificial intelligence with edge computing." *Proceedings of the IEEE* 107.8 (2019): 1738–1762.

Chapter 4

# Communication technologies for edge IoT

*Vishal Choudhary and Seema Kumari*

## 4.1 Introduction

In Edge Internet of things (IoT), the data processing and storage is done in close proximity, and it depends on the communication networks to transfer the data in between IoT devices, central storage, and processing system [1]. Using the data processing and analysis techniques near the data generation points makes the system efficient and fast. Miniaturization of hardware devices along with increasing their processing capabilities make edge IoT realities for general applications. In edge IoT, edge devices play a crucial role. These devices are equipped with sensors that capture data by sensing the environment in which they are deployed. There are uses of edge gateways as intermediate devices that process data from multiple IoT devices and transfer them over the network. Edge gateways have capabilities for further processing of data before transferring them to a central server. For deeper analysis and storage, edge IoT is dependent on the cloud platform. The main advantages of edge IoT are reduced latency, lower bandwidth, increased security along with improvement in reliability of an application. Edge IoT plays a major role in smart city applications, healthcare, and retail systems, as shown in Figures 4.1 and 4.2.

## 4.2 Integration of Wi-Fi with edge computing

The integration of Wireless Fidelity (Wi-Fi) with edge computing brings about significant improvements in latency, bandwidth usage, security, scalability, and user experience. It supports the efficient deployment and operation of modern and future technologies, making it a crucial aspect of network and application development [2]. Some of the key benefits are as follows.

### *4.2.1 High-speed connectivity*

Wi-Fi provides high-speed connectivity to edge computing through a combination of advanced technologies and optimized protocols designed to ensure efficient data transfer, low latency, and high throughput. Wi-Fi 6 (802.11ax)

DOI: 10.1201/9781003512066-4

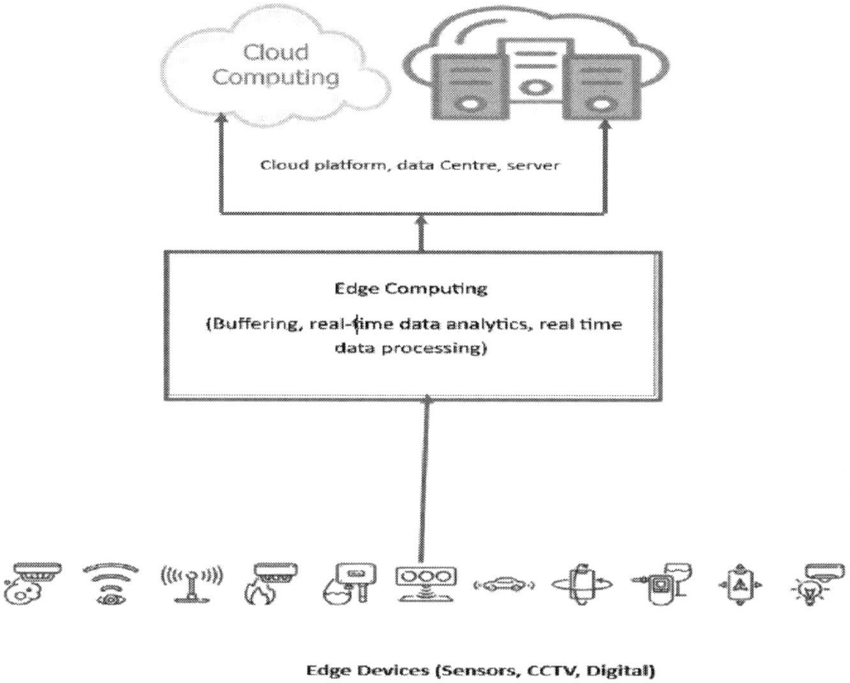


*Figure 4.1* Architectural View.

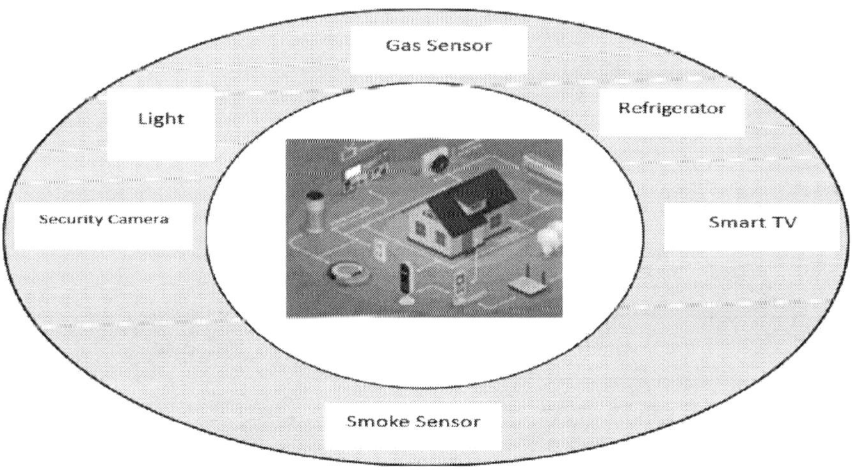


*Figure 4.2* Smart Home.

offers data rates up to 9.6 Gigabits per second (Gbps), which is essential for handling the large volumes of data generated by edge devices. Orthogonal Frequency Division Multiple Access (OFDMA) allows multiple devices to share channels simultaneously, increasing efficiency and reducing latency, while Multi-user MIMO (MU-MIMO) enables multiple devices to receive data simultaneously, optimizing throughput and connection quality. Wi-Fi 6 (sixth generation) also supports wide channel bandwidths, with 80 Megahertz (MHz) and 160 Megahertz (MHz) channels that can carry more data, essential for high-speed connectivity in edge computing environments where data are continuously generated and processed [3]. By using higher-order Quadrature Amplitude Modulation (1024-QAM), Wi-Fi 6 can encode more data within the same channel bandwidth, increasing the overall data rate. Beamforming directs Wi-Fi signals toward specific devices, enhancing signal strength and reliability, which is critical for stable and high-speed connections to edge devices. OFDMA and Target Wake Time (TWT) allow efficient use of the spectrum by assigning time slots for different devices and scheduling communication times, reducing contention and latency. The 5 Gigahertz (GHz) and 6 GHz bands offer more channels with less interference compared to the 2.4 GHz band, supporting higher data rates and more reliable connections necessary for edge computing, with extended Wi-Fi 6 (Wi-Fi 6E) into the 6 GHz band providing additional spectrum for higher data rates and reduced interference. Quality of Service (QoS) features prioritize critical edge computing tasks, ensuring they receive sufficient bandwidth and low latency, crucial for real-time processing. Mesh Wi-Fi systems use multiple nodes to create a unified network, providing consistent high-speed coverage throughout large areas, ensuring edge devices remain connected to the nearest node with optimal performance. Modern Wi-Fi routers are equipped with powerful hardware that can handle high-speed data transfers and complex routing required for edge computing, with regular firmware updates ensuring routers support the latest technologies and security features, maintaining high performance and reliability. Wi-Fi Protected Access 3 (WPA3) encryption provides robust security for data transmitted between edge devices and Wi-Fi access points, preventing unauthorized access and ensuring data integrity.

### 4.2.2 Local data processing

Wi-Fi provides local data processing to edge computing through several key mechanisms that optimize data transfer, processing efficiency, and network performance. Wi-Fi 6 (802.11ax), offering high data rates up to 9.6 Gbps, ensures rapid transmission of large volumes of data to and from edge computing nodes for real-time processing. Wi-Fi networks connect a wide range of edge devices, such as sensors, cameras, and Internet of Things (IoT) devices, to local edge servers, reducing the need to send large amounts of raw data to centralized cloud servers [4]. Technologies like OFDMA and MU-MIMO enhance spectrum usage, allowing multiple devices to share channels simultaneously, facilitating smooth local processing. Additionally, by enabling data processing at the edge, Wi-Fi significantly reduces latency, with features like TWT scheduling communication times to reduce contention. Wi-Fi networks support local storage solutions for data caching and processing near the source, crucial for applications requiring immediate responses. Advanced security protocols like WPA3 ensure secure data transmission between edge devices and local processing units, reducing risks associated with transmitting sensitive data to the cloud. Mesh Wi-Fi systems provide seamless and robust coverage across large areas, ensuring reliable connectivity for continuous data processing. Quality of Service (QoS) features prioritize critical data streams, ensuring essential edge computing tasks receive the necessary bandwidth and low latency. Channel bonding and higher-order modulation techniques, such as 1024-QAM, increase data capacity, allowing more local processing and conserving bandwidth. Modern Wi-Fi routers, equipped with powerful processors and enhanced memory, handle complex routing and high-speed data transfers, providing immediate computational resources. Overall, Wi-Fi supports local data processing in edge computing by delivering high-speed, low-latency connectivity, efficient data transmission technologies, local storage and processing, enhanced security, and mesh networking, facilitating real-time data analysis and decision-making at the edge, crucial for applications requiring immediate responses and high data throughput.

### 4.2.3 Scalability

Edge computing decentralizes data processing by handling computations at or near the source of data generation, reducing dependency on centralized cloud servers and allowing for more distributed and scalable data management [5]. By processing data locally, edge computing minimizes the need for large-scale data transfers, freeing up bandwidth and server resources for other tasks. By distributing computational tasks across multiple edge devices and servers, edge computing effectively balances the load, preventing any single point from becoming a bottleneck and enabling the system to handle increased workloads more efficiently. Edge computing can dynamically allocate resources based on current demand, allowing edge infrastructure to adapt to varying workloads and ensure optimal performance during peak usage times. Local data storage capabilities reduce the need to constantly send data back and forth to central data centers, enabling edge devices to handle large volumes of data without overloading central storage systems. By processing data closer to the source, edge computing reduces latency, ensuring that the system can scale effectively without compromising performance even as the number of edge devices and data volume increase. Edge computing architectures are often modular, allowing for easy addition or removal of components, enabling seamless scalability as new edge nodes can be added to the network without significant reconfiguration. By offloading data processing and analysis to the edge, edge computing reduces the volume of data transmitted over the network, conserving bandwidth and reducing congestion. Proximity to data sources allows for efficient data handling and quick scalability, ensuring that the system can scale horizontally by simply adding more edge nodes near the data sources. Edge computing works in conjunction with cloud services to provide a hybrid model where critical data are processed at the edge, and less time-sensitive data are handled in the cloud, optimizing resource utilization. Scalability in edge computing also comes from cost efficiency, as it lowers operational costs associated with data transfer and centralized processing. The ability to deploy cost-effective edge devices as needed provides a scalable and economical approach to managing growing data and processing demands. Overall, edge computing provides scalability by decentralizing data processing, enabling load balancing, offering elastic resource allocation, utilizing local data storage, reducing latency, featuring a modular architecture, enhancing network efficiency, and ensuring interoperability with cloud services, as well as cost efficiency, collectively ensuring that edge computing can effectively scale to meet increasing demands without compromising performance or reliability.

### *4.2.4 Cost-effectiveness*

Wi-Fi provides cost-effectiveness to edge computing through several mechanisms that reduce infrastructure, operational, and maintenance costs while optimizing resource utilization. Using Wi-Fi eliminates the need for extensive cabling and wired network infrastructure, which can be expensive to install and maintain [6]. Wireless networks are easier and cheaper to deploy, especially in environments where physical cabling is impractical or cost-prohibitive. This reduction in infrastructure costs makes it more affordable to set up and expand edge computing environments. Wi-Fi networks offer greater flexibility in deploying edge devices. Without the constraints of wired connections, edge devices can be placed in optimal locations to improve performance and coverage, reducing the costs associated with reconfiguring or expanding wired networks to accommodate new devices. Wi-Fi networks can easily scale to accommodate more devices without significant additional costs. Adding new edge devices to a Wi-Fi network is straightforward and does not require major investments in new infrastructure. This scalability ensures that the edge computing environment can grow to meet increased demand without incurring substantial expenses. Wi-Fi technologies, such as OFDMA and MU-MIMO, optimize the use of available bandwidth and improve network efficiency. By allowing multiple devices to share channels and receive data simultaneously, Wi-Fi ensures efficient data transmission and reduces the need for additional network resources, lowering operational costs. Modern Wi-Fi standards include features designed to reduce power consumption, such as TWT, which schedules communication times and allows devices to enter low-power states when not in use. These energy-saving features reduce the power consumption of edge devices and network infrastructure, leading to lower energy costs. Wi-Fi networks are generally easier to manage and maintain compared with wired networks. Wireless network management tools and technologies simplify the monitoring, configuration, and troubleshooting of Wi-Fi networks. Reduced maintenance requirements and simplified management translate into lower operational costs. Wi-Fi is widely adopted and compatible with a vast array of devices and existing network infrastructure. This compatibility allows organizations to leverage their current investments in Wi-Fi technology without needing to invest heavily in new equipment or systems, thereby reducing costs. Reliable Wi-Fi connectivity ensures that edge devices can communicate and process data efficiently, minimizing downtime and disruptions. Improved productivity and reduced downtime translate into cost savings, as less time and resources are spent on resolving connectivity issues. Upgrading Wi-Fi networks to newer standards, such as Wi-Fi 6, is often more cost-effective than overhauling the wired network infrastructure. Wi-Fi upgrades can provide significant performance improvements with relatively lower investment, ensuring that the edge computing environment remains efficient and cost-effective. Advanced Wi-Fi security protocols, like WPA3, provide robust security for data transmission, reducing the risk of costly data breaches and unauthorized access. By ensuring secure communication between edge devices and local processing units, Wi-Fi helps protect sensitive data and reduces potential financial losses associated with security incidents. In summary, Wi-Fi provides cost-effectiveness to edge computing by reducing infrastructure and operational costs, offering flexible and scalable deployment options, optimizing resource utilization, and lowering energy consumption. Its compatibility with existing infrastructure, ease of management, and robust security features further contribute to cost savings, making Wi-Fi an economical choice for supporting edge computing environments.

### *4.2.5 Problems associated with integration of Wi-Fi with edge computing*

Integrating Wi-Fi with edge computing offers significant benefits but also introduces several challenges that must be addressed for optimal performance and seamless operation. First, Wi-Fi networks can experience latency issues, particularly in crowded environments or areas with interference, which can impact the real-time processing capabilities required by edge computing applications. Security is another critical concern as decentralized edge environments increase the attack surface, necessitating robust measures to safeguard data integrity and privacy over Wi-Fi connections [7]. Scalability of Mobile Point-of-Sale (POS) systems poses a challenge when expanding Wi-Fi networks to accommodate growing numbers of edge devices while maintaining compatibility across diverse hardware and software platforms. Ensuring consistent QoS for latency-sensitive tasks is essential but challenging in dynamic edge environments where prioritizing traffic effectively is crucial. Moreover, managing power consumption efficiently is vital to extend the battery life of edge devices reliant on Wi-Fi connectivity. Effective management and orchestration of Wi-Fi networks across distributed edge locations require centralized control and monitoring capabilities, necessitating robust management tools and automation. Lastly, compliance with regulatory standards governing Wi-Fi usage and data privacy adds complexity, requiring adherence to regional and industry-specific regulations to integrate Wi-Fi networks securely into edge computing environments while ensuring legal compliance. Addressing these challenges comprehensively is essential to realizing the full potential of Wi-Fi-integrated edge computing for diverse IoT, industrial, and consumer applications.

#### *4.2.5.1 Interference and congestion*

Interference and congestion of Mobile POS pose significant challenges in Wi-Fi-based edge computing environments, impacting overall performance and reliability. Wi-Fi operates in the unlicensed spectrum, where various devices like Bluetooth and microwaves can introduce interference. In densely populated or industrial areas with numerous wireless devices, this interference can degrade signal quality, leading to packet loss and increased latency for edge devices [8]. Such issues can severely hinder the real-time processing capabilities crucial for edge computing applications. Additionally, high traffic and network congestion occur when multiple devices compete for limited bandwidth within Wi-Fi networks. In edge computing scenarios where multiple edge devices simultaneously transmit data to local servers or the cloud, congestion can cause delays and reduce responsiveness, affecting critical applications requiring timely data processing. These challenges collectively impact Wi-Fi network performance, lowering data transfer rates and reliability. Mitigation strategies include careful frequency planning to minimize interference, implementing QoS policies to prioritize essential traffic, utilizing wider channel bandwidths for increased throughput, managing network traffic effectively, and deploying advanced antenna technologies to enhance signal strength. By implementing these strategies, Wi-Fi can provide robust connectivity in edge computing deployments, ensuring reliable data transmission to support latency-sensitive applications.

#### *4.2.5.2 Security concerns*

Security concerns associated with Wi-Fi in edge computing environments are substantial due to the decentralized nature of data processing and inherent vulnerabilities in wireless communication. Ensuring data integrity and confidentiality is paramount as Wi-Fi networks transmit sensitive information between edge devices, local processing units, and centralized cloud services. Without robust encryption protocols like WPA3, there's a risk of data interception or alteration, particularly critical in industries such as healthcare and finance. Authentication mechanisms must be robust to prevent unauthorized access, as weak protocols or misconfigured controls could compromise operational security. Edge devices, often with limited processing power and irregular security updates, are susceptible to exploits, making them potential entry points for attackers aiming to infiltrate broader networks or launch malicious activities [9]. Man-in-the-Middle (MitM) attacks pose another significant threat, exploiting Wi-Fi's susceptibility to intercept and manipulate data exchanged between devices, especially concerning environments with numerous interconnected devices. Physical security of Wi-Fi access points and edge devices is crucial to prevent tampering or theft, which could compromise network integrity and data confidentiality [10]. Compliance with regulatory requirements is essential, ensuring edge computing deployments adhere to industry-specific standards and data protection laws governing Wi-Fi usage to avoid legal consequences and reputational damage. Continuous monitoring and rapid incident response are critical, enabling organizations to detect anomalies promptly, mitigate threats in real time, and uphold network security [11]. Addressing these challenges necessitates a holistic approach encompassing strong encryption, rigorous access controls, regular firmware updates, thorough security audits, and comprehensive staff training. By prioritizing these measures, organizations can effectively mitigate risks associated with Wi-Fi in edge computing environments, safeguarding data integrity and confidentiality across their operations [12].

#### *4.2.5.3 Range limitations*

The range limitations of Wi-Fi present substantial challenges when integrating with edge computing environments, affecting connectivity, reliability, and overall operational efficiency. Wi-Fi signals experience attenuation and degradation over distance and through physical obstacles like walls and floors, posing difficulties in maintaining consistent connectivity across expansive or remote edge environments. Devices positioned at the periphery of Wi-Fi coverage may suffer from reduced data transfer speeds, increased latency, or intermittent connectivity issues, which are critical drawbacks for real-time data processing needs. Moreover, operating in the unlicensed spectrum makes Wi-Fi susceptible to interference from other wireless devices and environmental factors such as weather conditions and physical obstructions. These interferences can lead to packet loss, errors, and reduced throughput, compromising the quality of service required for edge computing applications. Deploying Wi-Fi in edge environments demands careful planning to strategically place access points (APs) and implement technologies like mesh networking to extend coverage and optimize signal distribution. Balancing extended coverage with efficient power management becomes essential, as higher transmit powers or additional APs can increase energy consumption, particularly challenging in remote or mobile edge deployments with limited power resources. Addressing these range limitations involves leveraging technologies like beamforming and mesh networking, alongside strategic AP placement and optimization, to enhance Wi-Fi reliability and resilience in supporting critical edge computing operations.

### *4.2.6 Use cases of Wi-Fi and edge computing*

Wi-Fi integrated with edge computing facilitates a variety of practical use cases across different industries, leveraging the strengths of both technologies to enhance efficiency, connectivity, and real-time processing capabilities [13]. The following are some compelling use cases.

#### *4.2.6.1 Smart homes*

Smart homes represent a compelling use case for integrating Wi-Fi with edge computing, revolutionizing traditional living spaces into interconnected environments that prioritize convenience, security, and energy efficiency. Wi-Fi acts as the backbone of smart home ecosystems, connecting a diverse array of devices including smart thermostats, lighting systems, security cameras, door locks, appliances, and entertainment systems. These devices communicate wirelessly with each other and with cloud services, enabling seamless automation and remote control via smartphones or voice assistants. Edge computing plays a pivotal role by processing data locally within smart homes instead of relying solely on distant cloud servers. This approach significantly reduces latency, ensuring quick and responsive handling of critical applications such as security monitoring and real-time video streaming. For instance, edge devices can analyze video feeds locally to detect motion or identify faces, triggering immediate alerts or actions without needing to access the cloud. Security and privacy are enhanced through edge computing, as sensitive data are processed locally, reducing the risk of data breaches or unauthorized access. Devices within smart homes can perform encryption, access control, and authentication locally, ensuring robust security measures are in place. This localized processing of data, such as analyzing video footage from security cameras for anomalies, preserves privacy while maintaining high security standards. Energy efficiency is another key benefit enabled by Wi-Fi and edge computing in smart homes. Devices like smart thermostats can optimize energy usage based on real-time data analysis of occupancy patterns and external conditions, thereby reducing energy consumption without compromising comfort. The ability to learn user preferences over time allows these devices to intelligently adapt their operations, further enhancing energy efficiency. Scalability and flexibility are facilitated by Wi-Fi networks within smart homes, supporting the addition of new devices and expanding automation capabilities. Edge computing ensures that local hubs or gateways can manage a variety of devices efficiently without overloading central networks or cloud servers. This scalability ensures that smart home systems can grow organically while maintaining dependable performance levels. The integration of Wi-Fi and edge computing also elevates the user experience in smart homes, offering intuitive control interfaces such as mobile apps or voice commands. Edge computing ensures swift response times for user commands and updates, fostering a seamless and interactive interaction with smart home devices. Personalized services are made possible through edge computing in smart homes, which analyze user behavior and preferences locally to tailor experiences accordingly. For instance, smart lighting systems can adjust brightness and color based on individual preferences stored locally, providing a customized environment for each household member. Reliability and redundancy are enhanced by distributing computing tasks across edge devices within the home. Local processing guarantees that smart home functions continue to operate even if internet connectivity is disrupted temporarily or cloud services are unavailable, ensuring uninterrupted automation and security monitoring capabilities.

In summary, the integration of Wi-Fi with edge computing transforms smart homes by enabling seamless connectivity, local data processing for enhanced responsiveness, improved security and privacy measures, energy-efficient operations, scalability, personalized user experiences, and heightened reliability. These advancements not only make smart homes more intuitive and efficient but also adapt to the evolving needs and preferences of homeowners, creating a connected living experience that prioritizes convenience and sustainability.

#### *4.2.6.2 Industrial automation*

Wi-Fi integrated with edge computing offers a transformative solution for enhancing industrial automation, streamlining processes, and optimizing efficiencies across manufacturing and industrial sectors. This integration enables a range of capabilities that are pivotal for modernizing industrial operations:

Real-time Monitoring and Control: Wi-Fi facilitates immediate communication between industrial machinery, sensors, and control systems, enabling rapid data exchange without the delays typical of cloud-based processing. Edge devices equipped with Wi-Fi can collect data from sensors on the factory floor, analyze it locally, and transmit actionable insights swiftly to control systems or operators. This capability supports quick decision-making, enhances production efficiency, and minimizes downtime by enabling timely adjustments and responses.

Wireless Connectivity Across Facilities: Wi-Fi networks provide flexible and scalable connectivity across industrial sites, fostering seamless communication between distributed edge devices, machinery, and centralized control systems. This connectivity supports mobility within the industrial environment, allowing mobile devices and automated systems like Automated Guided Vehicles (AGVs) to maintain continuous communication with the central hub or other edge devices. Such mobility enhances operational flexibility, facilitates real-time data exchange, and adapts quickly to changing production demands. Edge Data Processing and Analytics: Edge computing empowers industrial automation by processing data locally at the edge—where it is generated—using Wi-Fi-connected devices. This approach minimizes the need to transmit large volumes of raw data to centralized servers, thereby reducing network congestion and latency. Localized data processing enables edge analytics to detect anomalies, predict equipment failures, optimize processes in real time, and enhance overall operational efficiency. These capabilities are crucial for industries aiming to improve productivity and responsiveness while leveraging data-driven insights for continuous improvement.

Scalability and Flexibility: Wi-Fi's inherent scalability supports the integration of additional edge devices and sensors into industrial automation networks with minimal infrastructure investment. New machinery, sensors, or production lines can easily connect to existing Wi-Fi networks, facilitating rapid deployment and scaling of operations. This flexibility enables industrial facilities to expand their capabilities, integrate emerging technologies seamlessly, and adapt swiftly to evolving manufacturing requirements without disrupting ongoing operations.

Remote Monitoring and Maintenance: Wi-Fi-enabled edge computing enables remote monitoring and predictive maintenance capabilities within industrial settings. By leveraging Wi-Fi for data transmission, engineers and maintenance teams can access real-time operational data, conduct diagnostics, and schedule maintenance tasks from any location within the facility or remotely. This proactive approach reduces downtime, extends equipment lifespan, and enhances overall equipment effectiveness (OEE), thereby optimizing operational efficiency and maintenance costs.

Enhanced Safety and Compliance: Wi-Fi networks support the deployment of advanced safety systems and compliance monitoring in industrial environments. Edge devices equipped with Wi-Fi can transmit safety-critical data in real time, such as temperature readings, gas levels, or equipment status, ensuring compliance with regulatory standards and mitigating risks associated with operational hazards. Enhanced safety measures contribute to maintaining a secure working environment for personnel while adhering to industry regulations.

Integration with IoT and Artificial Intelligence Technologies: Wi-Fi-connected edge devices in industrial automation facilitate seamless integration with IoT devices and artificial intelligence (AI) technologies. IoT sensors capture data on machine performance, environmental conditions, and production metrics, which edge computing processes locally to derive actionable insights [14]. AI algorithms deployed at the edge analyze data patterns, optimize production processes, and enable predictive maintenance, enhancing operational efficiency, productivity, and decision-making capabilities.

Cost Efficiency and Return on Investments: Wi-Fi integrated with edge computing offers cost-effective alternatives to traditional wired networks and cloud-based data processing models in industrial automation. By reducing infrastructure costs, optimizing resource utilization, and minimizing downtime through proactive maintenance, these solutions improve return on investment (ROI) for industrial automation initiatives. The scalability and flexibility of Wi-Fi networks also support future-proofing industrial operations against technological advancements and business growth, ensuring long-term sustainability and competitiveness in the manufacturing landscape. In summary, Wi-Fi integrated with edge computing revolutionizes industrial automation by enabling real-time monitoring and control, wireless connectivity across facilities, localized data processing and analytics, scalability, remote monitoring and maintenance capabilities, enhanced safety and compliance, integration with IoT and AI technologies, and cost efficiency. These capabilities empower industrial enterprises to achieve higher operational efficiency, improve productivity, ensure regulatory compliance, and drive sustainable growth in a competitive manufacturing environment.

#### *4.2.6.3 Healthcare*

Wi-Fi integrated with edge computing presents a transformative use case in healthcare, revolutionizing patient care, operational efficiency, and clinical workflows [15]. Here's an exploration of its application in this critical industry:

Real-time Patient Monitoring: Wi-Fi enables continuous and real time monitoring of patients' vital signs, such as heart rate, blood pressure, and oxygen levels, using wearable devices or sensors embedded in medical equipment. Edge computing processes these data locally, allowing healthcare providers to receive immediate alerts for abnormal readings and respond swiftly to critical situations without relying solely on centralized servers. This capability enhances patient safety, reduces response times, and supports proactive healthcare interventions.

Mobile Point-of-Care (POC) Solutions: Wi-Fi networks facilitate seamless connectivity between mobile devices (e.g., tablets, smartphones) used by healthcare professionals and centralized electronic health record (EHR) systems or diagnostic tools. Edge computing enables these devices to securely access and update patient information in real time, even in areas with limited or intermittent internet connectivity. Clinicians can review patient histories, update treatment plans, and access medical resources instantly at the point of care, improving efficiency and accuracy in decision-making.

Telemedicine and Remote Consultations: Wi-Fi-enabled edge devices support telemedicine initiatives by facilitating high-quality audio and video communications between patients and healthcare providers. Edge computing processes video streams locally, ensuring low latency and high-quality transmission, which is crucial for virtual consultations, remote monitoring of chronic conditions, and telehealth initiatives. This capability extends healthcare access to remote or underserved areas, enhances patient engagement, and reduces the need for in-person visits, thereby optimizing healthcare delivery and operational efficiency.

Medical Device Connectivity and Management: Wi-Fi connectivity in edge computing environments enables seamless integration and management of medical devices, such as infusion pumps, ventilators, and imaging equipment. Edge devices can monitor device status, perform diagnostics, and transmit operational data securely to centralized systems or cloud platforms for analysis. This connectivity improves device efficiency, ensures compliance with regulatory standards, and supports predictive maintenance to prevent equipment failures, thereby enhancing patient safety and reducing downtime.

Data Security and Privacy: Edge computing enhances data security in healthcare by processing sensitive patient information locally, within secure environments like hospitals or clinics. Wi-Fi networks encrypt data transmission between edge devices and central servers, ensuring compliance with healthcare privacy regulations (e.g., Health Insurance Portability and Accountability Act in the United States). By minimizing data exposure to external threats and unauthorized access, edge computing protects patient confidentiality and maintains trust in healthcare services.

Emergency Response and Disaster Management: Wi-Fi-enabled edge devices support emergency response teams and disaster management efforts by providing reliable communication and real-time data processing capabilities in crisis situations. Edge computing can analyze data from wearable sensors, monitor environmental conditions, and coordinate emergency responses efficiently, even when traditional communication infrastructures are disrupted. This resilience improves response times, enhances coordination among healthcare providers, and facilitates rapid decision-making to save lives during emergencies.

Healthcare Facility Optimization: Wi-Fi and edge computing enable smart healthcare facility management by integrating with building automation systems (BAS) and IoT devices. Edge devices can monitor energy consumption, optimize Heating, Ventilation, and Air Conditioning (HVAC) systems, and manage facility resources based on real-time data analytics. This integration enhances operational efficiency, reduces operational costs, and ensures a comfortable and safe environment for patients and healthcare staff.

In summary, Wi-Fi integrated with edge computing transforms healthcare delivery by enabling real-time patient monitoring, mobile POC solutions, telemedicine capabilities, secure medical device connectivity, data security, emergency response readiness, and healthcare facility optimization. These advancements enhance patient outcomes, streamline clinical workflows, improve operational efficiency, and support healthcare providers in delivering high-quality, accessible care across diverse healthcare settings.

#### *4.2.6.4 Retail*

Wi-Fi integrated with edge computing revolutionizes retail operations by offering a range of transformative applications that enhance customer experiences, optimize operational efficiency, and foster innovative business strategies. Through personalized customer engagement, retailers leverage Wi-Fi-enabled edge devices to gather real-time data on customer movements and behaviors within stores. These data allow for personalized offers and recommendations delivered directly to customers' mobile devices, thereby boosting engagement and influencing purchasing decisions in the moment. Moreover, edge computing combined with Wi-Fi supports real-time inventory management, utilizing sensors and Radio-Frequency Identification (RFID) tags to track stock levels and item locations seamlessly. By processing these data locally, edge devices ensure accurate inventory insights without dependence on centralized servers, minimizing stockouts and enhancing operational efficiency. In-store analytics further benefit from Wi-Fi connectivity, enabling retailers to analyze foot traffic patterns, dwell times, and customer behavior trends. Edge computing processes these data locally to generate actionable insights, empowering retailers to optimize store layouts, product placements, and marketing strategies based on real-time customer preferences and behaviors. Efficient checkout and payment processing are streamlined through Wi-Fi-connected self-service kiosks and mobile POS systems, which process transactions locally to reduce latency and ensure swift payment processing. This capability supports diverse payment methods and integrates loyalty programs in real time, enhancing customer satisfaction and operational flow. Supply chain management in retail is also optimized with Wi-Fi-enabled edge devices monitoring shipments, tracking goods, and providing timely updates on delivery statuses. This visibility enhances logistics efficiency, reduces shipping delays, and enables proactive decision-making to meet customer demands effectively. Interactive digital signage powered by Wi-Fi-connected edge devices transforms customer engagement within retail environments, offering personalized recommendations and promotional content based on customer preferences. Meanwhile, enhanced store security is achieved through Wi-Fi-enabled video monitoring systems employing edge computing for real-time video analytics. This ensures swift detection of suspicious activities, compliance with safety protocols, and improved incident response times, thereby enhancing overall store security and customer safety. Furthermore, offering guests Wi-Fi services in retail stores facilitates the collection of valuable customer insights through Wi-Fi analytics. These data provide demographic information, visitor trends, and detailed customer preferences, empowering retailers to refine personalized marketing campaigns, loyalty programs, and retention strategies. By processing these data locally, edge computing protects customer privacy while delivering actionable analytics for strategic decision-making.

In conclusion, Wi-Fi integrated with edge computing fundamentally transforms retail operations by enhancing customer engagement, optimizing inventory management, enabling real-time analytics, streamlining checkout processes, improving supply chain efficiency, bolstering store security, and providing invaluable customer insights. These advancements empower retailers to deliver personalized experiences, drive operational efficiencies, and achieve sustainable growth in a competitive retail landscape.

#### *4.2.6.5 Smart cities*

Wi-Fi integrated with edge computing represents a pivotal advancement in the evolution of smart cities, revolutionizing urban living through enhanced efficiency, sustainability, and quality of life across diverse domains. Real-time data processing and analysis are facilitated by Wi-Fi-connected edge devices scattered throughout city infrastructures. These devices gather data from sensors embedded in traffic lights, environmental monitors, and smart meters, allowing for immediate insights into traffic patterns, air quality, and energy consumption. Edge computing processes these data locally, minimizing latency and empowering city authorities to make timely decisions, such as dynamically managing traffic flow or responding swiftly to environmental concerns. Furthermore, edge computing enhances public safety and security by leveraging Wi-Fi-enabled surveillance systems and IoT devices. Local processing of video feeds from cameras and sensors enables real-time monitoring of public spaces, traffic incidents, and emergencies. Edge analytics can detect anomalies, identify suspicious activities, and promptly alert law enforcement or emergency services, thereby bolstering crime prevention and improving incident response times. Efficient management of urban infrastructure is streamlined by Wi-Fi and edge computing technologies. Edge devices equipped with Wi-Fi connectivity monitor the performance of street lighting, waste management systems, and water distribution networks in real time. They detect faults, leaks, or operational anomalies autonomously, enabling proactive maintenance and minimizing service downtime. This proactive approach not only reduces operational costs but also enhances overall reliability and sustainability of city services. Wi-Fi networks integrated with edge computing foster connectivity and digital inclusion by providing robust internet access through public Wi-Fi hotspots. These access points, powered by edge devices, bridge the digital divide by offering high-speed connectivity to residents and businesses in urban areas. This connectivity supports smart applications such as remote education, telemedicine, and e-government services, promoting economic development and social equity within the community. Smart mobility solutions are optimized through the combination of edge computing and Wi-Fi, transforming transportation systems and enhancing commuter experiences. Connected vehicles, traffic management systems, and smart parking solutions leverage real-time data processing to mitigate traffic congestion, reduce travel times, and improve road safety. Edge devices enable predictive maintenance of transportation infrastructure, ensuring reliable operation and minimizing disruptions for commuters. Environmental monitoring and sustainability efforts are bolstered by Wi-Fi-enabled edge devices within smart cities. Sensors measure air quality, noise levels, and waste management metrics, with edge computing processing these data locally. Monitoring trends and identifying pollution hotspots enable targeted interventions for cleaner air, reduced noise pollution, and efficient waste management practices, contributing to a healthier urban environment. Moreover, Wi-Fi integrated with edge computing optimizes energy management across smart city infrastructures, facilitating real-time monitoring and control of electricity usage, load balancing, and integration of renewable energy sources. This capability enhances energy efficiency, reduces carbon footprints, and supports sustainable urban development goals, ensuring cities operate with minimal environmental impact. Citizen engagement and quality of life are significantly enhanced by Wi-Fi and edge computing technologies. Interactive platforms and mobile applications powered by edge devices provide real-time information on public services, events, and community initiatives. Citizens can engage actively in city governance, provide feedback, report issues, and contribute to improving their neighborhoods. This fosters a sense of community and civic engagement, enhancing overall quality of life in smart cities.

In conclusion, Wi-Fi integrated with edge computing transforms smart cities by enabling real-time data processing, enhancing public safety, optimizing infrastructure management, improving connectivity and digital inclusion, advancing smart mobility solutions, promoting environmental sustainability, optimizing energy usage, and empowering citizen engagement. These advancements pave the way for more livable, resilient, and sustainable urban environments that cater to the evolving needs of residents, businesses, and visitors alike.

## 4.3 Integration of edge computing with bluetooth and bluetooth low energy

Integrating edge computing with Bluetooth and Bluetooth Low Energy (BLE) combines the advantages of decentralized processing and energy-efficient wireless communication, enabling a range of applications in IoT, smart environments, and real-time data processing [16]. Edge computing refers to processing data closer to the data source rather than relying on a centralized cloud, which reduces latency, bandwidth use, and enhances data security. Edge devices capable of processing data locally, such as gateways, routers, or powerful IoT devices, play a crucial role. Bluetooth is a wireless technology standard for exchanging data over short distances, commonly used in personal devices such as smartphones, headphones, and laptops. BLE, a variant of Bluetooth designed for low power consumption, is ideal for battery-powered IoT devices like health monitors, fitness trackers, smart locks, and beacons. Integration strategies include deploying BLE sensors to collect environmental data and using Bluetooth-enabled edge gateways to receive, process, and filter these data before sending critical information to the cloud. Local decision-making involves edge devices performing real-time analytics on data from BLE devices, which is essential for applications like smart buildings, autonomous vehicles, and industrial automation due to reduced latency. BLE's low power consumption ensures battery-operated devices can function for extended periods without frequent recharging, and efficient data handling at the edge reduces data transmission and conserves energy. For security and privacy, local data processing and robust encryption and authentication mechanisms are implemented to protect data during transmission between BLE devices and edge gateways. Scalability is achieved through a modular architecture, allowing easy addition or removal of BLE devices and edge gateways, and ensuring interoperability between different manufacturers and standards. Applications include smart homes and buildings for environmental monitoring and occupancy detection, healthcare for remote monitoring and fitness tracking, industrial IoT (IIoT) for predictive maintenance and asset tracking, and retail for personalized marketing and inventory management. Implementation considerations involve selecting appropriate hardware, developing software for edge devices, ensuring robust network infrastructure, and effective data management practices. Integrating edge computing with Bluetooth and BLE offers a powerful solution for real-time, efficient, and secure data processing across various domains, enhancing operational efficiency, reducing costs, and improving user experiences.

### 4.3.1 Benefits of integrating edge computing with bluetooth and BLE

Integrating edge computing with Bluetooth and BLE offers numerous benefits, enhancing the capabilities and efficiency of various applications. The following are the key benefits.

#### 4.3.1.1 Reduced latency

By processing data closer to the source, edge computing minimizes the time taken to analyze and respond to data inputs. This is crucial for real-time applications such as autonomous vehicles, industrial automation, and smart building systems where immediate decision-making is required.

#### 4.3.1.2 *Enhanced data security and privacy*

Local processing of sensitive data reduces the risk of data breaches during transmission to centralized cloud servers. This is particularly important in healthcare, financial services, and other sectors where data privacy is paramount [17]. Additionally, robust encryption and authentication mechanisms between BLE devices and edge gateways ensure secure communication.

#### 4.3.1.3 *Lower bandwidth usage and costs*

By filtering and processing data at the edge, only relevant information is sent to the cloud, significantly reducing the amount of data transmitted. This leads to lower bandwidth usage and associated costs, making it more cost-effective for organizations to manage their data infrastructure.

#### 4.3.1.4 *Energy efficiency*

BLE's low power consumption ensures that battery-operated devices can function for extended periods without frequent recharging. This energy efficiency is especially beneficial for remote or hard-to-access areas, where changing batteries can be challenging and costly.

#### 4.3.1.5 *Scalability and flexibility*

A modular architecture allows for easy scaling by adding or removing BLE devices and edge gateways based on the evolving needs of the application. This scalability is essential for growing IoT deployments and ensures that the system can adapt to new requirements without significant overhauls.

#### 4.3.1.6 *Improved reliability and resilience*

Local data processing at the edge reduces dependency on cloud infrastructure, which can be prone to outages and connectivity issues. This enhances the reliability and resilience of the system, ensuring continuous operation even when the cloud connection is unstable.

#### 4.3.1.7 *Real-time data analytics*

Edge devices can perform real-time analytics on data received from BLE sensors, enabling immediate insights and actions. This is vital for applications like smart cities, industrial monitoring, and healthcare, where timely data analysis can lead to proactive measures and improved outcomes.

#### 4.3.1.8 *Context-aware services*

Integrating BLE with edge computing allows for context-aware services that adapt based on real-time data. For example, in a smart building, BLE sensors can detect occupancy and adjust lighting or HVAC systems accordingly, optimizing energy use and enhancing user comfort.

#### 4.3.1.9 *Enhanced user experience*

By enabling quick responses and personalized interactions, the integration of edge computing with BLE enhances the overall user experience. For instance, retail applications can use BLE beacons to offer personalized promotions to customers in real time, increasing engagement and satisfaction.

#### 4.3.1.10 *Cost savings*

Reduced data transmission, lower bandwidth usage, and decreased reliance on cloud infrastructure contribute to overall cost savings. Additionally, the extended battery life of BLE devices and the ability to perform local processing can lower maintenance and operational costs.

### 4.3.2 *Challenges of integrating edge computing with bluetooth and BLE*

Integrating edge computing with Bluetooth and BLE presents several challenges, despite its numerous benefits. Addressing these challenges is crucial for successful deployment and operation of such systems. The following are the key challenges.

#### 4.3.2.1 *Security and privacy concerns*

- **Data Security:** Ensuring robust security measures to protect data during transmission between BLE devices and edge gateways is challenging. BLE's limited processing power can make it difficult to implement strong encryption and authentication mechanisms [18].
- **Local Data Storage:** Storing sensitive data locally on edge devices can pose security risks if the devices are compromised.

#### 4.3.2.2 *Interoperability issues*

- **Device Compatibility:** BLE devices from different manufacturers may have compatibility issues, making seamless integration difficult.
- **Standards and Protocols:** Variations in communication protocols and standards can hinder interoperability between different BLE devices and edge computing platforms.

#### 4.3.2.3 *Network and connectivity challenges*

- **Limited Range:** BLE has a limited communication range, which can be a problem in large-scale deployments or when devices are spread over a wide area.
- **Interference:** BLE operates in the 2.4 Gigahertz (GHz) frequency band, which is shared with other wireless technologies like Wi-Fi. This can lead to interference and degraded performance.

#### 4.3.2.4 *Power management*

- **Battery Life:** Although BLE is designed for low power consumption, managing the power usage of a large number of BLE devices to ensure long battery life can be challenging.
- **Energy Efficiency of Edge Devices:** Ensuring that edge devices, which process data locally, also operate efficiently without consuming excessive power is crucial [19].

#### 4.3.2.5 *Data management*

- **Data Overload:** Edge devices may face challenges in processing and storing large volumes of data generated by numerous BLE sensors.
- **Data Filtering:** Effectively filtering and processing data locally to ensure only relevant information is transmitted to the cloud can be complex.

#### 4.3.2.6 *Scalability*

- **System Expansion:** Scaling the system to accommodate an increasing number of BLE devices and edge gateways without compromising performance can be difficult.
- **Resource Allocation:** Efficiently managing computational and storage resources on edge devices to handle growing data volumes and processing demands is another challenge.

#### 4.3.2.7 *Latency and performance*

- **Real-Time Processing:** Ensuring that edge devices can perform real time data analytics and decision-making without significant delays is significant.
- **Network Latency:** Minimizing latency in communication between BLE devices and edge gateways, and between edge gateways and the cloud, is critical for real-time applications [20].

#### 4.3.2.8 *Deployment and maintenance*

- **Installation Complexity:** Deploying a large number of BLE devices and edge gateways in diverse environments can be logistically complex.

- **Maintenance Challenges:** Regular maintenance of a distributed network of edge devices and BLE sensors, including firmware updates and troubleshooting, can be resource-intensive.

##### *4.3.2.9 Cost considerations*

- **Initial Investment:** The upfront cost of purchasing and deploying BLE devices and edge gateways can be significant.
- **Ongoing Expenses:** Continuous maintenance, power consumption, and potential upgrades add to the long-term costs of the system.

##### *4.3.2.10 Regulatory and compliance issues*

- **Compliance with Standards:** Ensuring that the integration of edge computing and BLE complies with industry standards and regulations, especially in sectors like healthcare and finance, is crucial.
- **Data Privacy Laws:** Adhering to data privacy laws and regulations, such as the EU General Data Protection Regulation, when processing and storing data locally on edge devices is essential.

## 4.4 Future trends in integration of edge computing with bluetooth and bluetooth low energy

The integration of edge computing with Bluetooth and BLE is poised to evolve significantly in the coming years. Several trends are emerging that will shape the future of this integration, enhancing its capabilities and expanding its applications. The following are the key future trends.

### *4.4.1 Advancements in BLE technology*

- **Increased Range and Speed:** Future iterations of BLE (e.g., Bluetooth 5.1 and beyond) will offer increased range and faster data transfer rates, making it more suitable for a wider range of applications, including those requiring higher bandwidth.
- **Enhanced Location Services:** Improved accuracy in location-based services using BLE will enable more precise indoor navigation, asset tracking, and geofencing applications.

### *4.4.2 AI and machine learning at the edge*

- **Intelligent Edge Devices:** The incorporation of AI and machine learning (ML) capabilities into edge devices will allow for more sophisticated data processing and analytics. This will enable predictive maintenance, anomaly detection, and real-time decision-making.

- **On-Device Learning:** Advances in on-device ML will allow edge devices to learn and adapt over time, improving their performance and efficiency without the need for constant cloud updates.

### 4.4.3 *Edge AI chips and hardware*

- **Specialized AI Chips:** The development of specialized AI chips for edge devices will enhance processing power while maintaining energy efficiency. These chips will enable more complex computations to be performed locally.
- **Compact and Efficient Hardware:** Innovations in hardware design will lead to more compact, energy-efficient edge devices capable of handling increased computational loads.

### 4.4.4 *Enhanced security measures*

- **Edge Security Solutions:** As security remains a critical concern, future trends will focus on developing advanced security solutions specifically designed for edge computing environments [21]. This includes hardware-based security modules, encrypted communication protocols, and AI-driven threat detection.
- **Blockchain Integration:** Blockchain technology could be integrated with edge computing to provide secure, tamper-proof data exchanges between BLE devices and edge nodes.

### 4.4.5 *Edge and cloud synergy*

- **Hybrid Architectures:** A more seamless integration between edge and cloud computing will emerge, where edge devices handle real-time processing and the cloud is used for deeper analytics and long-term storage. This synergy will optimize resource use and improve overall system efficiency.
- **Dynamic Workload Distribution:** Intelligent systems will dynamically distribute workloads between edge and cloud based on real-time requirements, ensuring optimal performance and resource utilization.

### 4.4.6 *IoT and fifth-generation network integration*

- **5G Connectivity:** The rollout of fifth-generation (5G) networks will complement BLE by providing ultra-low latency and high-speed connectivity, facilitating faster and more reliable communication between edge devices and the cloud [22].
- **Massive IoT Deployments:** With 5G, it will be possible to support a massive number of connected BLE devices, enabling large-scale IoT applications such as smart cities and industrial automation.

### 4.4.7 Edge-to-edge communication

- **Inter-Edge Collaboration:** Future systems will enable edge devices to communicate directly with each other, bypassing the cloud. This will facilitate faster data sharing and collective decision-making in distributed environments.
- **Mesh Networking:** BLE mesh networking will become more prevalent, allowing BLE devices to form robust, self-healing networks that enhance coverage and reliability.

### 4.4.8 Sustainability and energy efficiency

- **Green Computing Initiatives:** Emphasis on sustainability will drive the development of more energy-efficient edge devices and BLE sensors. This includes advancements in low-power processing and renewable energy-powered edge nodes.
- **Battery Innovations:** Breakthroughs in battery technology will extend the life of BLE devices, reducing the frequency of replacements and maintenance.

### 4.4.9 Standardization and interoperability

- **Unified Standards:** Efforts to create unified standards for edge computing and BLE will enhance interoperability between devices from different manufacturers, simplifying integration and deployment.
- **Open-Source Platforms:** The rise of open-source edge computing platforms will foster innovation and collaboration, enabling developers to build and deploy edge applications more efficiently.

### 4.4.10 New applications and use cases

- **Smart Cities:** The integration of edge computing with BLE will drive the development of smart city applications, including intelligent traffic management, environmental monitoring, and public safety.
- **Healthcare Innovations:** Advanced healthcare applications will emerge, leveraging real-time data processing and AI to provide personalized medicine, remote diagnostics, and improved patient care.
- **IIoT:** Edge computing and BLE will enable more sophisticated IIoT applications, such as advanced robotics, automated quality control, and efficient supply chain management.

## 4.5 Integration of edge computing with zigbee

Integrating edge computing with Zigbee, a low-power, low-data-rate wireless communication technology, can significantly enhance the capabilities of

IoT systems. Zigbee is widely used for applications requiring secure, reliable wireless communication over short distances, such as home automation, smart energy, and industrial control systems. The following is an overview of how these technologies can be integrated and what benefits they offer.

### 4.5.1 Data collection and processing

- **Zigbee Sensors:** Deploy Zigbee sensors to collect data from the environment (e.g., temperature, humidity, motion). These sensors can transmit data to edge devices.
- **Edge Gateways:** Use Zigbee-enabled edge gateways to receive data from Zigbee sensors. The gateway processes the data locally, performs initial analysis, and filters out noise before sending critical information to the cloud.

### 4.5.2 Local decision-making

- **Real-Time Analytics:** Edge devices can perform real-time analytics on data received from Zigbee devices. For example, in a smart building, an edge gateway can adjust lighting or HVAC systems based on occupancy data from Zigbee sensors.

### 4.5.3 Reduced latency

Local processing ensures decisions are made quickly without the latency involved in cloud communication, which is crucial for applications like autonomous vehicles or industrial automation.

### 4.5.4 Energy efficiency

- **Optimized Communication:** Zigbee's low power consumption ensures that battery-operated devices can operate for extended periods without frequent recharging. This is especially beneficial in remote or hard-to-access areas.
- **Efficient Data Handling:** By processing data at the edge, only relevant data are sent to the cloud, reducing the amount of data transmitted and conserving energy.

### 4.5.5 Security and privacy

- **Local Data Processing:** Sensitive data can be processed locally at the edge, reducing the risk of data breaches during transmission to the cloud.
- **Encryption and Authentication:** Implement robust encryption and authentication mechanisms for data transmitted between Zigbee devices and edge gateways to ensure secure communication.

### 4.5.6 *Scalability*

- **Modular Architecture:** Design a modular architecture where Zigbee devices and edge gateways can be easily added or removed based on the requirements. This scalability is essential for growing IoT deployments.
- **Interoperability:** Ensure interoperability between different Zigbee devices and edge gateways to accommodate various manufacturers and standards.

### 4.5.7 *Challenges of integrating edge computing with zigbee*

Integrating edge computing with Zigbee presents several challenges that need to be addressed for successful deployment and operation, as follows.

#### *4.5.7.1 Interference and range limitations*

Zigbee operates in the 2.4 GHz frequency band, which is shared with other wireless technologies like Wi-Fi and Bluetooth. This can lead to interference issues, especially in dense deployment scenarios. Additionally, Zigbee's range may be limited, requiring careful planning and potentially the use of repeaters in larger deployments.

#### *4.5.7.2 Scalability*

While Zigbee supports mesh networking for extending the range and improving reliability, managing a large number of devices in a Zigbee network can be complex. Ensuring efficient communication and resource management becomes crucial as the network scales up.

#### *4.5.7.3 Power management*

Zigbee devices are typically designed for low power consumption, but managing the power requirements across numerous devices in edge computing setups can still be challenging. Balancing power efficiency with the need for continuous operation is essential, particularly in battery-powered devices.

#### *4.5.7.4 Data handling and processing*

Edge devices in Zigbee networks may face challenges in processing and managing large volumes of data generated by connected devices. Efficient data aggregation, filtering, and processing at the edge are necessary to optimize bandwidth and ensure timely response to data insights.

#### *4.5.7.5 Security concerns*

Ensuring robust security measures within Zigbee networks is critical to protect data integrity and prevent unauthorized access. Zigbee devices may need to implement encryption, authentication, and secure communication protocols to mitigate potential security risks.

#### *4.5.7.6 Integration with existing systems*

Integrating Zigbee networks with existing IT infrastructure or other communication protocols can be complex due to compatibility issues and differing standards. Ensuring seamless interoperability and data exchange between Zigbee devices and edge computing platforms is essential for cohesive system operation.

#### *4.5.7.7 Deployment and maintenance*

Deploying Zigbee networks across diverse environments requires careful planning to address installation complexities and ensure optimal device placement for reliable connectivity. Additionally, ongoing maintenance of Zigbee devices and networks, including firmware updates and troubleshooting, is necessary to sustain performance and security.

Addressing these challenges involves leveraging advanced network management tools, implementing robust security protocols, optimizing power management strategies, and ensuring compatibility and scalability in Zigbee-based edge computing solutions. By overcoming these obstacles, organizations can harness the benefits of Zigbee's low-power, reliable communication for efficient edge computing deployments across various industries.

## 4.6 Integration of edge computing with cellular networks

Integrating edge computing with cellular networks represents a significant advancement in enhancing network capabilities and enabling new applications. This integration offers several key benefits. First, edge computing reduces latency by processing data closer to its source, minimizing the time data travel between devices and the cloud via cellular networks. This capability is essential for real-time applications such as autonomous vehicles, industrial automation, and augmented reality. Second, cellular networks provide wide coverage and scalability, supporting large-scale deployments of edge computing solutions across diverse geographical locations. The evolution to 5G networks further enhances this with higher bandwidth, enabling faster data transmission speeds for applications like high-definition video streaming, remote surgery, and large-scale data analytics. Additionally, cellular networks ensure robust connectivity with high reliability and resilience, crucial for mission-critical applications in healthcare, finance, and public safety. Security measures such as encryption and authentication protocols protect data in transit between edge devices and cloud servers, addressing cybersecurity concerns. Integrating edge computing with cellular networks allows for a hybrid approach where edge devices handle real-time processing locally, optimizing resource utilization and improving system efficiency. This flexibility supports various deployment models in remote areas, smart factories, or urban environments, ensuring consistent performance and connectivity. Leveraging existing cellular infrastructure also reduces additional network investment costs, while edge computing minimizes data transfer expenses by processing and filtering data locally before transmitting relevant information to the cloud. Despite these advantages, challenges such as managing network congestion, ensuring interoperability across different network technologies, and complying with regulatory requirements need collaborative efforts from telecom providers, edge solution developers, and regulatory bodies to achieve seamless integration and effective utilization of edge computing capabilities across cellular networks [23].

### 4.6.1 Benefits of integrating edge computing with cellular networks

Integrating edge computing with cellular networks offers a multitude of benefits that enhance network capabilities and enable innovative applications across various industries. The following are the key advantages.

#### 4.6.1.1 Low latency for real-time applications

Edge computing reduces latency by processing data closer to where it is generated, minimizing delays in data transmission between edge devices and the cloud via cellular networks. This capability is critical for real-time applications such as autonomous vehicles, industrial automation, and augmented reality, where split-second decisions are essential.

#### 4.6.1.2 Scalability and wide coverage

Cellular networks provide extensive coverage and scalability, making them suitable for large-scale deployments of edge computing solutions. Edge devices can seamlessly connect across diverse geographical locations, facilitating expansive IoT deployments and smart city initiatives without compromising connectivity.

#### 4.6.1.3 High bandwidth support

The evolution to 5G networks significantly increases cellular bandwidth, enabling faster data transmission speeds and supporting bandwidth-intensive applications. This enhancement is crucial for applications like high-definition video streaming, remote surgery, and large-scale data analytics, which require robust and high-speed connectivity.

#### *4.6.1.4 Reliability and resilience*

Cellular networks are designed to deliver robust connectivity with high reliability and resilience. They ensure continuous operation even in challenging environments or during peak usage periods, providing dependable connectivity for mission-critical applications in healthcare, finance, and public safety sectors.

#### *4.6.1.5 Enhanced security measures*

Cellular networks implement stringent security protocols, including encryption and authentication mechanisms, to protect data transmitted between edge devices and cloud servers. This enhances data privacy and safeguards against cybersecurity threats, ensuring secure data transmission and storage.

#### *4.6.1.6 Optimized resource utilization*

Integrating edge computing with cellular networks allows for a hybrid architecture where edge devices perform real-time processing locally. This approach optimizes resource utilization by reducing the dependency on cloud resources for immediate decision-making, thereby improving overall system efficiency and performance.

#### *4.6.1.7 Flexibility in deployment*

Edge computing integrated with cellular networks supports flexible deployment models across diverse environments, including remote locations, smart factories, and urban areas. This flexibility ensures consistent performance and connectivity, adapting to various industry-specific use cases and operational requirements.

#### *4.6.1.8 Cost efficiency*

Leveraging existing cellular infrastructure minimizes the need for additional network investments. Edge computing further reduces operational costs by processing and filtering data locally before transmitting relevant information to the cloud. This approach lowers data transfer expenses and optimizes operational efficiencies, contributing to cost savings over time.

### **4.6.2 Challenges of integrating edge computing with cellular networks**

Integrating edge computing with cellular networks presents several challenges that need to be addressed to ensure successful deployment and operation. The following are the key challenges.

#### 4.6.2.1 *Network congestion and bandwidth management*

Cellular networks can experience congestion, especially in densely populated areas or during peak usage times. Managing bandwidth effectively becomes crucial to maintain reliable connectivity and ensure consistent performance for edge computing applications.

#### 4.6.2.2 *Latency concerns*

While edge computing reduces latency by processing data closer to the point of generation, latency can still be a concern in cellular networks due to varying network conditions and distances between edge devices and cloud servers. Minimizing latency is essential for real-time applications like autonomous vehicles and industrial automation.

#### 4.6.2.3 *Reliability and resilience*

Cellular networks are generally reliable, but they can face disruptions due to network outages, coverage APs, or environmental factors. Ensuring high availability and resilience of connectivity is critical, especially for mission-critical applications in healthcare, finance, and public safety.

#### 4.6.2.4 *Security and privacy*

Securing data transmitted between edge devices and cloud servers over cellular networks is paramount. Cellular networks use encryption and authentication mechanisms, but vulnerabilities can still exist. Implementing robust security measures to protect sensitive data from potential cyber threats is essential.

#### 4.6.2.5 *Interoperability across network generations*

Integrating edge computing across different cellular network generations (e.g., 2G/3G/4G/5G) can be challenging due to differences in protocols and capabilities. Ensuring seamless interoperability and backward compatibility is necessary to support diverse devices and applications.

#### 4.6.2.6 *Cost of infrastructure and deployment*

Deploying edge computing infrastructure across cellular networks involves costs associated with hardware, software, and network upgrades. Managing these upfront and ongoing costs while achieving ROI requires careful planning and investment.

#### *4.6.2.7 Regulatory compliance*

Compliance with regulatory requirements, such as data protection laws (e.g., GDPR), is essential when processing and transmitting data over cellular networks. Adhering to industry standards and regulations ensures legal and ethical use of data in edge computing applications.

#### *4.6.2.8 Complexity of management and maintenance*

Managing a distributed network of edge devices and ensuring their maintenance, including firmware updates and troubleshooting, can be complex and resource intensive. Establishing effective management practices is crucial to optimize performance and longevity of edge computing deployments.

## 4.7 Integration of edge computing with satellite communication

Integrating edge computing with satellite communication extends computational capabilities to remote and mobile environments globally. Satellite networks offer ubiquitous coverage, supporting edge applications in sectors like maritime, aviation, disaster response, and remote industrial operations. This integration reduces latency for real-time applications such as autonomous vehicles and telemedicine, leveraging satellite's low-latency data transmission. Satellites also provide resilience against network disruptions, ensuring continuous connectivity in harsh environments. Scalable and secure, satellite networks employ robust encryption, though challenges include high costs, latency management, and technical complexities requiring collaboration among stakeholders for effective deployment and optimization.

### *4.7.1 Benefits of integrating edge computing with satellite communication*

Integrating edge computing with satellite communication offers several significant benefits, enhancing connectivity and computational capabilities in remote and mobile environments globally. The following are the key advantages:

#### *4.7.1.1 Global coverage*

Satellite communication provides ubiquitous coverage, reaching remote and underserved areas where terrestrial networks are unavailable or impractical. This enables edge computing applications in sectors such as maritime, aviation, disaster response, and remote industrial operations.

#### 4.7.1.2 *Low latency connectivity*

Edge computing reduces latency by processing data closer to the point of origin. When coupled with satellite communication, which offers low-latency data transmission despite geographical distances, real-time applications like autonomous vehicles, remote monitoring, and telemedicine can operate effectively.

#### 4.7.1.3 *Resilience and redundancy*

Satellites offer inherent resilience against terrestrial network disruptions, making them ideal for applications requiring continuous connectivity in harsh or disaster-prone environments. This redundancy ensures reliability crucial for mission-critical operations.

#### 4.7.1.4 *Scalability*

Satellite networks support scalable deployments across vast geographic regions without the need for extensive terrestrial infrastructure. This scalability is beneficial for IoT deployments, environmental monitoring, and global supply chain management.

#### 4.7.1.5 *Security and privacy*

Secure data transmission between edge devices and cloud servers via satellite communication is ensured through robust encryption and authentication protocols. This addresses cybersecurity concerns in remote operations.

### 4.7.2 Challenges of integrating edge computing with satellite communication

Integrating edge computing with satellite communication offers transformative potential but also comes with several challenges that need to be addressed for successful deployment, as follows.

#### 4.7.2.1 *Latency*

Despite advancements, satellite communication introduces higher latency compared to terrestrial networks. This can impact real-time applications that require low-latency responses, such as autonomous vehicles and telemedicine.

#### 4.7.2.2 *Cost*

Satellite communication infrastructure and services can be costly, particularly for high-bandwidth applications. Managing costs while ensuring a viable ROI remains a significant challenge.

#### 4.7.2.3 *Complexity*

Integrating edge computing with satellite networks involves technical complexities. These include interoperability between different satellite constellations, synchronization issues, and managing communication between edge devices and satellites.

#### 4.7.2.4 *Reliability*

While satellites offer resilience against terrestrial network disruptions, they can still be affected by weather conditions and orbital dynamics. Ensuring continuous and reliable connectivity is crucial for mission-critical applications.

#### 4.7.2.5 *Security*

Securing data transmitted via satellite communication is essential. Robust encryption and authentication mechanisms must be implemented to protect sensitive information against potential threats.

#### 4.7.2.6 *Regulatory compliance*

Compliance with international regulations governing satellite communication, spectrum allocation, and data privacy such as at mobile POS systems poses additional challenges. Adhering to these regulations while deploying edge computing solutions is essential to avoid legal and operational issues.

## 4.8 Conclusion

The integration of edge computing with Wi-Fi), Bluetooth/BLE, Zigbee, cellular networks, and satellite communication marks a significant advancement in enhancing network capabilities and enabling diverse applications across industries. Wi-Fi's high-speed connectivity and local data processing capabilities make it ideal for real-time applications in smart homes, industrial automation, healthcare, retail, and smart cities. However, challenges such as interference, security concerns, and range limitations must be addressed to ensure seamless integration.

Bluetooth and BLE offer benefits like low power consumption and com patibility with a wide range of devices, yet interoperability issues and data overload pose challenges. Future trends, including enhanced capabilities and integration with AI, promise advancements in these areas. Zigbee's low power consumption and mesh networking capabilities are advantageous for scalable IoT deployments but face challenges such as limited data rate and interference.

Integrating edge computing with cellular networks enhances connectivity with wide coverage, high data rates, and low latency critical for applications requiring mobility and reliability. Challenges like cost management, power consumption, and network congestion need careful consideration to optimize performance. Satellite communication extends global coverage and resilience but introduces challenges such as high latency, cost, and integration complexity.

In conclusion, addressing these challenges through technological innovation, strategic planning, and regulatory compliance is essential to fully leverage the benefits of integrating edge computing with these wireless technologies. The future holds promise with ongoing advancements in each area, paving the way for transformative applications in diverse sectors globally.

## References

[1] W. Yu, F. Liang, X. He, et al., "A Survey on the Edge Computing for the Internet of Things," *EEE Access,* vol. 6, pp. 6900–6919, 2018.
[2] W. Rafique, "Complementing Internet of Things (IoT) Services Through Software Defined Networking and Edge Computing: A Comprehensive Survey," *IEEE Communications Surveys & Tutorials*, vol. 22, no. 3, pp. 1761–1804, 2020.
[3] C. Deng, "IEEE 802.11 Be Wireless Fidelity (Wi-Fi)7: New Challenges and Opportunities," *IEEE Communications Surveys & Tutorials*, vol. 22, no. 4, pp. 2136–2166, 2020.
[4] M. Silva, "The Path Towards Virtualized Wireless Communications: A Survey and Research Challenges," *Journal of Network and Systems Management*, vol. 32, no. 12, 2023.
[5] P. J. Escamilla-Ambrosio, "Distributing Computing in the Internet of Things: Cloud, Fog and Edge Computing Overview," in *NEO 2016. Studies in Computational Intelligence*, Springer, Cham, 2018, pp. 87–115.
[6] M. Talebkhah, "Edge Computing: Architecture, Applications and Future Perspectives," in *2020 IEEE 2nd International Conference on Artificial Intelligence in Engineering and Technology (IICAIET)*, Kota Kinabalu, 2020.
[7] S. Szott, "Wi-Fi Meets ML: A Survey on Improving IEEE 802.11 Performance With Machine Learning," *IEEE Communications Surveys & Tutorials*, vol. 24, no. 3, pp. 1843–1893, 2022.
[8] R. S. Alonso, "A Survey on Software-Defined Networks and Edge Computing over Internet of Things (IoT)," in *Highlights of Practical Applications of Survivable Agents and Multi-Agent Systems. The PAAMS Collection. PAAMS 2019*, Springer, 2019.
[9] V. Choudhary, "The Highly Secure Polynomial Pool-Based Key Pre-Distribution Scheme for Wireless Sensor Network," *Journal of Discrete Mathematical Sciences and Cryptography*, vol. 23, no. 1, pp. 95–114, 2020.
[10] V. Choudhary, "A Distributed Key Management Protocol for Wireless Sensor Network," in *Advanced Informatics for Computing Research. ICAICR 2018*, Singapore, 2018.
[11] V. Choudhary, T. Taruna, and L. B. Purbey, "A Comparative Analysis of Cryptographic Keys and Security," in *3rd International Conference and Workshops on Recent Advances and Innovations in Engineering (ICRAIE)*, Jaipur, 2018.
[12] R. S. Vishal, "Efficient Cluster Key Management Scheme for Wireless Sensor Networks," in *Artificial Intelligence and Sustainable Computing. ICSISCET 2022*, Gwalior, Madhya Pradesh, 2023.
[13] L. U. Khan, "Edge-Computing-Enabled Smart Cities: A Comprehensive Survey," *IEEE Internet of Things*, vol. 7, no. 10, pp. 10200–10232, 2020.

[14] T. D. Vass, "Internet of Things (IoT) in Supply Chain Management: A Narrative on Retail Sector Sustainability," *International Journal of Logistics Research and Applications*, vol. 24, no. 6, pp. 605–624, 2021.
[15] G. K. Walia, "AI-Empowered Fog/Edge Resource Management for Internet of Things (IoT) Applications: A Comprehensive Review, Research Challenges, and Future Perspectives," *IEEE Communications Surveys & Tutorials*, vol. 26, no. 1, pp. 619–669, 2024.
[16] A. Nikoukar, "Low-Power Wireless for the Internet of Things: Standards and Applications," *IEEE Access*, vol. 6, pp. 67893–67926, 2018.
[17] H. Habibzadeh, "A Survey on Cybersecurity, Data Privacy, and Policy Issues in Cyber-Physical System Deployments in Smart Cities," *Sustainable Cities and Society*, vol. 50, 2019.
[18] P. K. Bharti, V. Choudhary, and S. Gupta, "Fuzzy Logic Based Routing Algorithm for Mobile Ad Hoc Networks," in *International Conference on High Performance Architecture and Grid Computing*, Chandigarh, 2011.
[19] S. T. Vishal Choudhary, "The Highly Secure Polynomial Pool-Based Key Pre-Distribution Scheme for Wireless Sensor Network," *Journal of Discrete Mathematical Sciences and Cryptography*, vol. 23, no. 1, pp. 95–114, 2020.
[20] S. T. Vishal Choudhary, "A Distributed Key Management Protocol for Wireless Sensor Network," in *International Conference on Advanced Informatics for Computing Research*, Shimla, 2019.
[21] V. Choudhary and S. Taruna, "An Intrusion Detection Technique Using Frequency Analysis for Wireless Sensor Network," in *2021 International Conference on Computing, Communication, and Intelligent Systems (ICCCIS)*, Greater Noida, 2021.
[22] V. Choudhary, D. S. Taruna, and L. B. Purbey, "A Comparative Analysis of Cryptographic Keys and Security," in *3rd International Conference and Workshops on Recent Advances and Innovations in Engineering (ICRAIE)*, Jaipur, 2018.
[23] S. T. Vishal Choudhary, "Fuzzy Analysis for Nodes Deployment Strategies in Wireless Sensor Network," *International Journal on Recent and Innovation Trends in Computing and Communication*, vol. 5, no. 6, pp. 852–855, 2017.

Chapter 5

# Advanced machine learning and artificial intelligence techniques at the edge

*Anish Chhetry, Rahul Kala, and Vaibhav Malviya*

## 5.1 Introduction

Over the last decade, Artificial Intelligence (AI) and Machine Learning (ML) have transitioned from emerging technologies to essential tools across numerous industries, including healthcare, finance, manufacturing, and consumer electronics. This rapid evolution has been fuelled by advances in computational power, the accessibility of huge datasets, and the advancement of complex algorithms. ML, in particular, revolves around the interaction between vast amounts of data and complex mathematical models that learn to recognise patterns, predict outcomes, and automate decision-making. However, the traditional approach to ML, which relies heavily on centralised cloud infrastructure, is increasingly being challenged by the rise of smart devices and the vast amounts of data generated at the network's edge.

Traditionally, datasets were collected, stored in large databases, and processed on cloud servers to train ML and Deep Learning (DL) models. This centralised approach allowed for the aggregation of massive datasets, which were essential for training models with high accuracy and generalisability. However, as the number of linked devices has grown exponentially, so too has the volume of data they produce. The explosion of data, much of it generated at the network's edge, has created significant challenges for traditional cloud-based ML models, particularly concerning data privacy, latency, and the cost of data transmission.

The emergence of the Internet of Things (IoT) has been a major driver of this trend. The term "Internet of Things" describes the network of physical objects that have sensors, software, and other technologies built into them so they can communicate and share information with other systems and devices connected online. The IoT has become a transformative force in various sectors, including healthcare, intelligent cities, industrial automation, agriculture, and monitoring of the environment. The widespread adoption of IoT has been driven by several factors, including the decreasing cost of sensors and communication technologies, wireless networking advancements, and the increasing availability of high-speed internet access. As a result, IoT

DOI: 10.1201/9781003512066-5

devices are now used almost everywhere, generating a continuous stream of data that requires analysis and processing to extract valuable insights.

However, the rapid development of IoT has also highlighted the limitations of traditional cloud-based architectures. IoT devices, often constrained by limited computing power, memory, and battery life, create enormous volumes of data that require processing and analyses to provide meaningful insights. While cloud computing offers the necessary resources for such large-scale data processing, the dependence on centralised cloud servers introduces several challenges. First, transmitting data from edge devices to distant cloud centres for processing can result in high latency, which is unacceptable for applications that need to make decisions in real time, like industrial robotics, autonomous cars, and remote healthcare monitoring. Second, the cost of transmitting massive amounts of data across a network can be excessively high, particularly in environments with limited bandwidth or high network traffic. Third, the transmission of sensitive data to centralised cloud servers raises significant privacy and security concerns, especially in sectors such as healthcare and finance, where data breaches can have severe consequences.

In response to these challenges, the concept of Edge Computing has gained traction as a viable solution for handling data nearer to its origin. Edge computing involves decentralising computation by placing processing power at the network's edge, near the devices that generate the data. This method reduces the requirement to transmit data to centralised cloud servers, thereby lowering latency, reducing bandwidth costs, and enhancing privacy and security. Edge Computing is particularly well suited for real-time processing applications, like autonomous systems, augmented reality, and intelligent city infrastructure.

Edge ML represents the convergence of edge computing and ML, where ML models are trained and deployed at the network's edge. By enabling ML computations to occur nearer to the data origin, Edge Machine Learning addresses several limitations of traditional cloud-based ML models. For example, by processing data locally on edge devices, Edge Machine Learning can significantly reduce the latency associated with data transmission, making it possible to achieve real-time decision-making. Moreover, Edge Machine Learning can reduce some of the computational burden from cloud servers, thereby optimising the use of cloud resources and reducing operational costs. Most importantly, by reducing the amount of private information that must be transmitted over the network, Edge Machine Learning improves data privacy by reducing the chance of data breaches.

Despite the advantages of Edge Machine Learning, deploying ML models at the edge presents its own set of challenges. Among the main obstacles, one is the limited computational power and memory available on edge devices, which are often constrained by hardware limitations. Training ML models from scratch requires enormous amounts of data and significant computational resources, which are typically not available on edge devices. As a result, traditional ML models trained in centralised cloud environments are often too enormous and complex to be deployed directly on edge devices.

To overcome these challenges, several innovative approaches have been proposed, including Federated Learning, Knowledge Distillation, and Transfer Learning. Medical applications constitute one of the largest use cases of modern ML, where hospitals and radiology labs are prohibited from sharing data due to privacy issues. Other use cases like AI in home and personal robotics face similar issues where privacy prohibits hosting vast amounts of collected data for learning centralised models. This gave rise to Federated Learning, where the learning is local across all devices, and thus, there is no loss of privacy. At the same time, the learned model is shared across all devices, commonly coordinated via a centralised system, and thus, the learned model is nearly as good as the one trained on the entire dataset. Apart from adding privacy, the system makes better use of hundreds of edge devices, and the learning computational requirements get parallelised for computational efficiency. Federated Learning has opened new avenues for learning that were previously restricted due to privacy laws.

Attributed to the advantages of DL, researchers are increasingly designing deeper and more complex networks to solve common ML problems. It is widely believed that deeper networks are over-regularised leading to complex architectures with limited actual representation capabilities, and thus many deeper architectures can be summarised into smaller architectures. The process of taking a deep and complex ML model and distilling its knowledge into a lower-complexity model is called Knowledge Distillation. The deeper model is called the teacher model, whose knowledge is distilled into a lower complexity model called student, while the student aims to have a minimal loss of accuracy compared to the teacher. On the one hand, distillation significantly compresses the model, allowing the model to be applied in near-real time settings for resource-constrained edge environments. On the other hand, smaller models are easier to analyse and interpret.

Many new problems may not have large datasets available, while DL approaches are generally data hungry. Humans can learn one task and transfer the learnings to another task, like learning Hindi makes it easier to learn Sanskrit since several language generics remain the same. Similarly, a machine trained on object recognition knows how to extract good edge, corner, and other features that make them good candidates to learn for related tasks on images such as Visual Place Recognition. Taking a machine trained on one task and fine-tuning the machine for a related task is called as Transfer Learning.

Artificial Intelligence of Things (AIoT) makes use of IoT devices and applies the algorithms and inference capabilities of the Artificial Intelligence algorithms. This has led to decentralised use of AI across thousands of IoT devices, where the devices capture, analyse, and reason about the data at the edge without requiring communication with the central server; while to enable the overall system to learn from all devices, the knowledge may be communicated and shared. The concept has applications in a variety of domains from medical to smart cities, where such AI-enabled devices result in more efficient systems.

Despite the potential of AIoT, several challenges must be addressed to fully realise its benefits. Deploying AI models on resource-limited devices requires innovative approaches to optimise computation and memory usage. Techniques such as model compression, quantisation, and pruning are essential to enabling AI models to run efficiently on edge devices with limited resources. Additionally, the dynamic and distributed nature of AIoT systems necessitates robust methods for managing computational resources across the edge cloud continuum. Security and privacy also remain significant concerns, as the decentralised nature of edge computing introduces new vulnerabilities that must be addressed to ensure the safe and reliable operation of AIoT systems [1–4].

This chapter offers a comprehensive exploration of the integration of AI with the IoT, which is supported by edge computing. This amalgamation enables the following key contributions:

(i) Fundamental of Things and Technologies: We depict an overview of the essential technologies underlying AIoT, including the architecture of IoT, advanced AI, and edge computing paradigms, along with the necessary hardware and systems.
(ii) Motivations for Combining the AI Tool with Edge Computing: We explore the motivations for integrating AI with edge computing, focusing on two main aspects: how AI algorithms can optimise edge computing and how edge computing enables faster, more stable AI applications by processing data closer to where they are generated.
(iii) Edge Computing Enhancement: We briefly discuss recent approaches and models to deploy AI training and inference tasks within the edge computing architecture, analysing their strengths and weaknesses. This analysis helps to understand how AI can be most effectively implemented in edge computing environments.
(iv) Advanced Applications: We discover applications of AIoT across diverse domains, such as healthcare, intelligent cities, industrial automation, and smart homes, highlighting the transformative potential in the era of AIoT.
(v) AI Inference and Decentralised Training: We also summarise the recent advances in AI inference across the AIoT hierarchy, from edge devices to cloud servers, and discuss technologies that enable decentralised AI training across end devices and edge servers.

## 5.2 Related works

We start the chapter by considering numerous recent studies and applications along with explaining their core functionalities. The technological brief of the background technologies and applications will be given in the subsequent sections. Numerous studies have explored the application of AI and ML, including DL, within the IoT domain. For instance, one study [5] highlights that combining AI with IoT not only simplifies human tasks but also introduces concerns related to security and ethics. Another study [6] reviews various ML techniques used in IoT applications across sectors like healthcare, smart grids, and vehicular communication, emphasising the core objectives of ML within the IoT. To address issues like high latency and data transmission costs, edge computing has been examined as a solution in the IoT context. One paper [7] shows that AI can make edge devices smarter and more efficient, enabling AI models to run directly at the network's edge.

Further research [8] investigates the potential applications of ML techniques in the areas of software-defined networking, edge computing, IoT applications, device identification, traffic analysis, and system security. Another comprehensive review [9] covers the characteristics of IoT data, advanced ML and DL methods, and various IoT applications that use Deep Neural Networks (DNNs). This review also examines methods and technologies that allow DL models to operate on devices with inadequate resources, such as edge servers. Additionally, another study [10] discusses techniques for compressing and accelerating DL algorithms, making them more appropriate for use on mobile and embedded devices, which are often resource limited. These techniques are essential for meeting the demands of real-time applications and guarantee user privacy. Another study [11] proposes a new offloading strategy that is crucial for improving the implementation of DL-based applications in IoT systems, especially those supported by edge computing and cloud infrastructure.

AI also plays a crucial role in effectively allocating resources in edge computing environments. For example, one study [12] introduces a dynamic approach to dealing with the runtime of sensing and data transmission processes in AI-powered IoT devices, particularly within industrial platforms. Another research effort [13] presents a deep reinforcement learning method that strategically orchestrates caching resources and edge computing to maximise system efficiency in vehicular networks. A survey [14] looks at how AI and edge computing are combined in the Internet of Vehicles, where AI is mainly utilised for task scheduling, resource allocation, and vehicle movement prediction in dynamic environments.

Interestingly, there is still a gap in research when it comes to establishing a unified concept and architecture for the AIoT. Most existing studies on AI implementation in IoT systems rely heavily on cloud platforms, which are not ideal for applications that require immediate responses. Edge computing addresses this limitation by increasing cloud capabilities to the network's edge.

## 5.3 Design philosophy

Edge AI, the practice of running AI models directly on edge devices, presents unique challenges due to limited resources and privacy concerns. To address these issues, various techniques have been developed to optimise AI performance at the edge. This section delves into the key categories of techniques: distributed/collaborative techniques, fine-tuning techniques, model compression techniques, traditional ML, DL, reinforcement learning, and other optimisation methods.

Each of these techniques offers distinct advantages in enhancing the efficiency, privacy, and performance of AI models deployed on edge devices. By understanding these approaches, researchers and practitioners can select the most suitable methods for their specific edge AI applications.

### 5.3.1 *Collaborative or distributed techniques*

This section will discuss various collaborative/distributed methods for training machine learning models directly at the edge. These methods address issues like communication bottlenecks and privacy concerns in edge computing.

#### 5.3.1.1 *Federated learning*

Federated Learning allows ML models to be trained on numerous edge devices without transferring private information to a central server. As an alternative, every device trains a different version of the model and only communicates updates to a central server, which aggregates them to produce the final model.

The key methods in Federated Learning include the following:

Federated Stochastic Gradient Descent and Federated Averaging: Common techniques for combining model updates [15].

$$\beta_{a+1} = \sum_{n=1}^{N} \frac{X_n}{X} \beta_{a+1}^{n} \tag{1}$$

where $\beta_{a+1}$ is a central model parameter, $N$ is the number of participants, $X_n$ is the number of samples of participant $n$, $X$ is the number of samples of all participants, and $\beta_{a+1}^{n}$ *is* the local model parameter of participant $n$.

Edge Federated Learning and Feature Dimension Sequence Labelling (FedSel): These are improvements that reduce computation and enhance data handling [16, 17].

Over-The-Air Federated Learning: This Optimises wireless communication for faster model updates. It minimises resource consumption by gathering the models "over the air" once [18].

Federated Learning faces challenges like dealing with diverse data types, malicious users, and communication issues. Solutions include advanced methods like Hybrid Federated Learning and Hierarchical Federated Learning to improve performance and privacy.

#### 5.3.1.2 *Split learning*

Split Learning is a method where an ML model is divided into parts, with each part trained on different devices. Only the results of the training (not the raw data) are shared between devices. Split Learning is good for privacy and works well in resource-limited settings, but it can be slower compared to Federated Learning [19].

#### 5.3.1.3 *Other collaborative learning methods*

Swarm Learning and Gossip Learning are other ways to train models collaboratively.

(i) Swarm Learning: Swarm Learning combines different techniques to work efficiently across many devices. It efficiently uses communication methods and signal processing to operate in real time [20].
(ii) Gossip Learning: Gossip Learning doesn't use a central server and relies on devices sharing updates directly with each other, which is useful for decentralised environments. It is more suitable for mobile devices [21].

### 5.3.2 *Fine-tuning techniques*

This section will discuss various fine-tuning methods for training ML models directly at the edge. These techniques help models efficiently learn and adapt on edge devices while conserving resources and protecting privacy.

#### 5.3.2.1 *Transfer learning*

Transfer Learning takes a model trained on one task and adapts it to a new but related task. This is useful when you don't have much data. It helps speed up training and works well on edge devices [22].

For example:

Tiny Transfer Learning: Makes learning efficient with limited memory.
RepNet: Adjusts models to be more memory efficient.
Mobile Transfer Learning: Optimises models for edge devices.
Federated Transfer Learning: Combines Transfer Learning with Federated Learning to protect privacy while adapting models.

#### 5.3.2.2 *Incremental learning*

Incremental Learning updates models with new data over time, without losing previously learned information. This method helps models continue improving as new data come in [23].

For example:

Learning with Sharing: Reduces memory use while maintaining accuracy.

Privacy-preserving Indoor Localization and Occupancy Tracking Engine: Recognises human activities on edge devices.

RILOD: Updates object detection models to recognise new objects without forgetting old ones.

#### 5.3.2.3 *Meta learning*

Meta Learning helps models learn new tasks quickly using minimal data by training them to adapt to various situations. This is especially useful on edge devices with limited resources [24].

For example:

Adaptation Aware Network Pruning: Makes models faster to adapt.

Continual Meta Learning: Learns from a sequence of tasks.

FedMC: Combines Meta Learning with Federated Learning for improved performance across devices.

### 5.3.3 *Model compression techniques*

This section will discuss different model compression methods for training ML models directly at the edge. These compression techniques make it easier and more efficient to train and deploy ML models on edge devices by reducing their size and computational demands.

#### 5.3.3.1 *Knowledge distillation*

Knowledge distillation is the process of educating a smaller, simpler model (the student) to behave similarly to a larger, more complex model (the teacher). This facilitates the usage of the smaller variant on edge devices [25].

For example:

Knowledge Distillation with Federated Learning: Combines Knowledge Distillation with Federated Learning to keep data private while simplifying models.

Knowledge Distillation for Edge Devices: Uses Knowledge Distillation to make models more efficient for edge devices, such as in recommendation systems or health monitoring systems.

#### *5.3.3.2 Quantisation*

By using fixed point or integer numbers instead of floating point ones, quantisation lowers the accuracy of the model's computations. As a result, models operate more quickly and consume less memory on edge devices [26].

There are two main types of quantisation, as follows:

(i) Quantisation Aware Training: Fine-tunes the model during training to handle the effects of reduced precision.
(ii) Post-Training Quantisation: Adjusts a pre-trained model to work with lower precision.

For example:

- Quantisation with Federated Learning: Reduces the size of models trained across multiple devices while maintaining data privacy.
- Quantisation for Edge Training: Uses quantisation to make on device training more efficient.

#### *5.3.3.3 Model pruning*

Model Pruning involves removing parts of a model, like certain weights or nodes, to make it smaller and faster. This can be done before or during training to make the model more suitable for edge devices [27].

For example:

Pruning with Federated Learning: Reduces model size to lower communication costs and training time in Federated Learning setups.

Pruning for Edge Applications: Uses pruning to improve the efficiency of models used for tasks like activity recognition or reinforcement learning.

### *5.3.4 ML algorithms*

ML algorithms, excluding DL and Reinforcement Learning, are categorised based on the availability of labelled data:

(i) Supervised Learning uses labelled data, with methods like support vector machines, boosting, and random forests.

(ii) Unsupervised Learning discovers patterns in unlabelled data, using techniques like K-means clustering and principal component analysis.
(iii) Semi-Supervised Learning combines both unlabelled and labelled data, with methods like label propagation and graphical models.

While conventional ML algorithms need complex feature engineering and may struggle with large datasets, they are energy efficient, cost effective, and easier to deploy. In edge computing, these algorithms can be tailored to match the resources and tasks of each device [28].

### 5.3.5 *Deep learning*

Like the human brain, DL is highly effective in extracting high-level features from unprocessed data for tasks like prediction and classification. Pooling, convolutional, and fully connected layers are among the several layers that make up DL models [11].

The key DL models include the following:

(i) Deep Neural Networks: Comprising several hidden layers, DNNs are powerful for learning complex patterns.
(ii) Convolutional Neural Networks: Convolutional neural networks are effective for image recognition and have applications in edge computing like fault detection.
(iii) Recurrent Neural Networks: Recurrent neural networks (RNNs) handle sequence data but have limitations, which Long Short-Term Memory networks address.

DL is particularly useful in edge computing for tasks like anomaly detection, task scheduling, and privacy protection, especially when ample labelled data are available.

### 5.3.6 *Reinforcement learning and deep reinforcement learning*

Reinforcement Learning involves interacting with the environment to train models, making it effective for decision-making tasks. Supervised ML algorithms need labels often obtained by a human annotator. Reinforcement Learning algorithms make an agent interact with an environment (often a simulated environment) that learns to maximise the cumulative sum of expected rewards. Reinforcement Learning algorithms can thus learn a navigation policy for the agent for sequential and complex problems [29]. A common Reinforcement Learning algorithm is Q-learning, which calculates rewards for actions in various states.

Deep Q-Networks enhance Q-learning by using deep neural networks to handle complex systems. Deep Q-Networks are particularly effective in edge computing environments but can face instability. To counter this, Deep Q-Networks incorporate an experience replay mechanism, which stabilises learning by utilising prior experiences [30].

### 5.3.7 Other optimisation techniques

#### 5.3.7.1 Binary neural networks

A neural network that uses binary values (-1 or 1) as weights and activations rather than complex numbers is known as a Binary Neural Network. Binary Neural Networks are very compact and require less computing power and memory, making them ideal for small, resource-limited devices [31].

For example, imagine having a super-simple calculator that only uses 1s and -1s instead of float numbers. This makes it much faster and easier on your phone or IoT device.

#### 5.3.7.2 Spiking neural networks

A Spiking Neural Network (SNN) is a neural network that mimics how real brains work by sending signals (spikes) between neurons. An SNN uses less energy because it only processes information when necessary, which is great for battery-powered devices [32].

For example, think of it as a smart light bulb that only turns on when it detects motion, saving energy compared to a light that is constantly on.

#### 5.3.7.3 Forward-forward algorithm

Forward-Forward algorithm is an alternative to the traditional back propagation method for training neural networks. Instead of one forward and one backward pass, it uses two forward passes. It reduces memory usage and computation, making it better suited for devices with limited resources [33].

For example, instead of taking a double-step approach to training a model, this method simplifies it to a single, efficient step that is easier to handle on small devices.

#### 5.3.7.4 Other techniques

(i) Data Booleanisation: This process converts data into a simple binary format to save energy and resources.
(ii) Generative Adversarial Networks: Generative Adversarial Networks (GNNs) consist of dual networks; a discriminator that determines

whether the data are authentic or fraudulent, and a generator that produces the data [34].

(iii) K-Nearest Neighbours: This algorithm classifies data points based on their proximity to other labelled points. ProtoNN is an optimised version of K-Nearest Neighbours, specifically designed for resource-constrained devices by reducing the training data size and projecting it into a lower-dimensional space [35].

**Manhattan Distance:** Manhattan Distance measures the total absolute difference between the coordinates of two points.

$$M_d = \sum_{k=1}^{x} |a_{1k} - a_{2k}| \tag{2}$$

**Euclidean Distance:** Euclidean Distance measures the straight-line distance between two points in Euclidean space.

$$E_d = \left( \sum_{k=1}^{x} (a_{1k} - a_{2k})^2 \right)^{\frac{1}{2}} \tag{3}$$

**Minkowski Distance:** Minkowski Distance is a generalised metric that includes both Euclidean and Manhattan distances as special cases, depending on the order parameter.

$$M_k = \left( \sum_{k=1}^{x} (a_{1k} - a_{2k})^p \right)^{\frac{1}{p}} \tag{4}$$

where $x$ is the distance and $a_1$ and $a_2$ are the points

(iv) Tree-Based ML Algorithms: These algorithms are commonly used for regression and classification but can be resource intensive. Bonsai is an emerging algorithm that adapts traditional tree algorithms for use on severely resource–limited IoT devices. It does this by learning a single sparse tree and making nonlinear predictions with minimal memory and processing power [36].

## 5.4 Problems in edge computing

There are three main problems in edge computing.

### 5.4.1 Computing offloading

Computing offloading is the process by which terminal devices with low processing power transfer some or all of their computations to servers with greater processing power. Initially introduced in cloud computing, this concept has been adapted for edge computing.

The main issue is deciding when and how much of the work to offload and which server to send the workload to.

### 5.4.2 Resource allocation

In edge computing, tasks are spread across different edge devices instead of being handled by a central cloud, which saves network bandwidth.

The main issue is managing and distributing these limited resources efficiently among edge devices, which is crucial for edge computing to work effectively.

### 5.4.3 Privacy and security

Protecting data isolation and security in edge computing is a major concern.

The main issue is edge computing's decentralised nature and the limited capability of edge devices makes it harder to apply traditional cloud security measures. New methods are required to ensure data safety in this environment.

## 5.5 Edge AI applications

In recent years, AI has achieved significant advancements across various domains like intelligent cities, intelligent manufacturing, and the Internet of Vehicles. These applications demand higher levels of network latency and stability. Unfortunately, regular cloud computing often falls short to meet these necessities. As a result, some researchers have begun utilising edge computing to provide computational and storage resources closer to the edge of the network. To highlight the benefits of edge computing in AI applications, this section focuses on summarising research findings related to its use in intelligent cities, intelligent manufacturing, and the Internet of Vehicles.

### 5.5.1 Intelligent city

Intelligent cities, and their increasing population density, focus on enhancing energy efficiency, reducing energy consumption, reducing traffic jams, and improving security and quality of life [37]. These cities rely on data generated by various devices, such as surveillance cameras, smartphones, and environmental sensors. AI is necessary for processing and analysing this vast amount of data. However, the intensive computing tasks required by AI pose problems in maintaining network stability and real-time response within a city environment. To solve this problem, some researchers have started working on edge computing.

#### *5.5.1.1 City security*

Intelligent cities require continuous monitoring of infrastructure and urban operations to guarantee residents' safety. AI plays an important role in improving the accuracy of security event identification, but the problem here is managing the massive amounts of data generated by these systems. Edge computing helps reduce network pressure by processing most data locally, reducing the amount of information that needs to be transmitted to the cloud. This approach enhances response speed and reduces latency in critical applications like traffic monitoring, facility monitoring, and crime detection.

For example, edge computing–enabled AI systems can quickly identify individuals in urban monitoring systems, making cities safer. The use of distributed AI algorithms across edge computing systems further improves response times and system efficiency, but now the problem lies in balancing AI complexity with the limited computing resources available on edge devices.

#### *5.5.1.2 Urban healthcare*

In intelligent cities, healthcare systems increasingly rely on IoT devices and cloud computing to collect and analyse personal medical data. Edge computing enhances these systems by reducing latency and ensuring stable data transmission, which is vital for applications like remote diagnosis, early disease detection, and infectious disease prevention.

For example, edge computing can preprocess data from wearable devices before it is uploaded to the cloud, enabling quicker responses and reducing network strain. Additionally, edge computing–based AI systems can improve the efficiency and accuracy of medical diagnoses, particularly in time-sensitive situations like early-stage lung cancer detection. Despite these advancements, challenges persist in developing AI algorithms that are both lightweight and accurate, as well as in managing energy consumption for wearable devices used in these healthcare applications.

#### *5.5.1.3 Urban energy management*

Urbanisation drives a significant increase in energy consumption, creating problems for efficient energy management in intelligent cities. Edge computing, combined with AI, offers solutions for real-time energy management by predicting energy consumption and optimising usage. This involves dynamic energy scheduling, where tasks are executed on the edge, with some coordination with cloud-based systems [38]. By deploying AI algorithms on the edge, cities can reduce energy waste and improve overall efficiency. A typical edge computing–based intelligent city power management system is shown in Figure 5.1.

However, the growing number of edge devices contributes to increased energy demand, which increases the global energy crisis. A possible solution is to power these devices with renewable energy, though this introduces challenges in ensuring consistent energy supply and balancing the energy needs of various edge devices within the city.

### 5.5.2 *Intelligent manufacturing*

Involving edge computing and AI into industrial production enhances efficiency, safety, and quality by optimising hardware usage and enabling distributed computing.

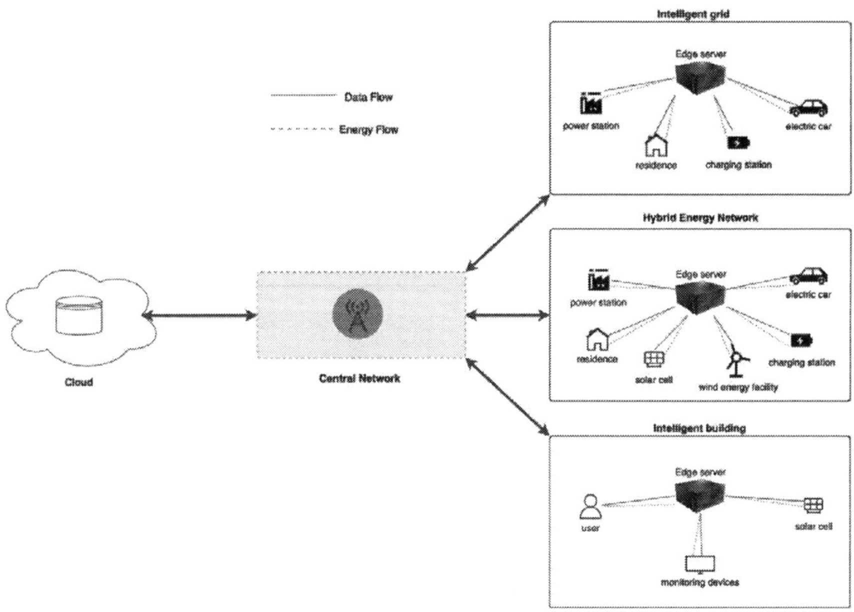


*Figure 5.1* A typical intelligent city power management system has three main parts:

1. Cloud: This is the central hub with strong control capabilities and strong computing resources.
2. Edge Servers: By analysing data near the point of collection, these manage local energy control.
3. Electrical Devices: These include various equipment and sensors used by people, energy producers, and consumers, all of which are connected to the system.

#### 5.5.2.1 *Dynamic control*

The iRobot Factory, an intelligent robot factory system was proposed that uses edge computing to adjust production lines in real time, process data, and apply AI for improved feedback control, reducing delays and enhancing cooperation among devices [39].

#### 5.5.2.2 *Asset monitoring*

Edge computing frameworks allow real-time monitoring and diagnostics of factory machinery, reducing delay and improving accuracy.

Non-intrusive systems are based on edge computing and cloud–based AI with Long Short-Term Memory algorithm to classify and manage equipment more effectively.

#### 5.5.2.3 *Product quality monitoring*

AI-driven systems identify defective products at the edge and cloud levels, increasing detection accuracy and efficiency while minimising data transmission costs.

#### 5.5.2.4 *Micro seismic monitoring*

In oil and gas production, edge computing and AI platforms process seismic data locally, improving real-time detection accuracy and reducing network strain, achieving over 96.00% accuracy and significant transmission efficiency of around 90.00%.

### 5.5.3 *Internet of vehicles*

The Internet of Vehicles (IV) is an emerging field that plays a significant role in advancing intelligent transportation and improving the driving experience. IV helps reduce traffic jam, lowers accident rates, and improves passenger comfort and safety through the integration of various intelligent systems, automotive applications, and road condition sensors.

Traditional cloud computing struggles to meet IVs' needs for steady networks and low latency due to its limitations. IVs' distributed nature requires more efficient resource allocation and task scheduling, which cloud computing alone cannot adequately provide.

Combining edge computing with AI offers quicker and more accurate control, better network communication, and improved user experiences. By processing data closer to the source, edge computing lowers latency and boosts efficiency. AI algorithms are increasingly used to optimise task offloading, resource allocation, and enhance vehicle intelligence. A typical edge computing–based IV structure is shown in Figure 5.2 [40].

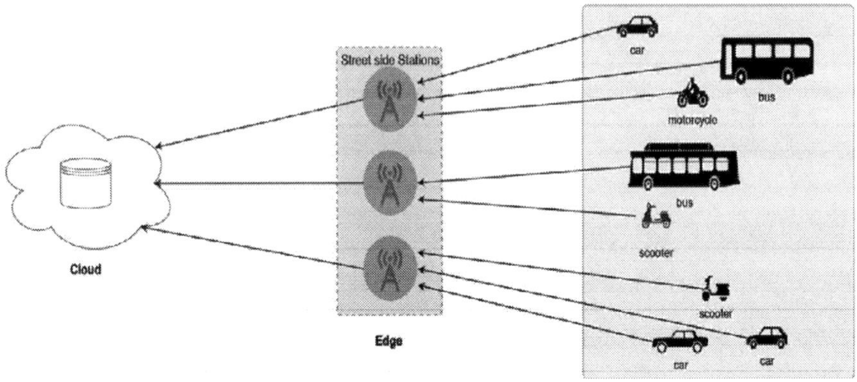


*Figure 5.2* A common structure of the Internet of Vehicles involves edge computing, where roadside units equipped with processing capabilities are utilised. In this setup, vehicles can offload computing tasks to these nearby roadside units rather than relying on distant cloud servers for processing.

#### 5.5.3.1 *Task and resource management*

Efficient task offloading and resource management are critical in IVs due to the dynamic nature of vehicular networks. Advanced algorithms, like ant colony optimisation and Deep Reinforcement Learning, are employed to address these challenges, taking into account data dependencies, vehicle movement, and network conditions.

#### 5.5.3.2 *Optimising onboard experience*

As autonomous driving technology evolves, the demand for automotive entertainment will grow. Edge computing–based caching strategies, combined with AI, help manage network content efficiently and provide a seamless user experience by predicting vehicle movements and optimising content delivery.

#### 5.5.3.3 *Enhancing vehicle intelligence*

AI under edge computing architecture enables vehicles to perform complex tasks such as automatic control, route analysis, and traffic prediction [41]. Various DL algorithms, like convolutional neural networks, long short-term memory, and restricted Boltzmann machines, are used to process data and make real-time decisions.

#### 5.5.3.4 *Challenges*

Despite the advancements, IVs faces challenges like maintaining stable communication while vehicles are in motion, ensuring data privacy, and creating incentive mechanisms for resource sharing among vehicles. Addressing these issues is essential for the widespread adoption of Internet of Vehicles.

## 5.6 Artificial intelligence inference for AIoT

Deep neural networks are classically designed to be complex and require large datasets to ensure high accuracy. However, performing AI inference in the cloud can introduce network delays, which can be detrimental for time-sensitive applications. Due to their size and computational demands, these models are often too resource intensive to be deployed directly on constrained end devices. To address this, one method is to simplify the models to reduce computational requirements, while another effective strategy is to offload more complex inference tasks to resource-rich edge nodes. The following discussion reviews various methods to optimise inference on devices, enable cooperative inference with edge nodes, and ensure privacy preserving techniques.

### 5.6.1 On device inference

Two primary methods are used to reduce the computational demands of AI models on devices:

#### 5.6.1.1 Network compression

Researchers aim to create smaller deep neural network models with fewer parameters but maintained accuracy. Notable models include Xception [42], MobileNets [43], YOLO [44], and SqueezeNet [45], which use techniques such as depth-wise separable convolutions and 1×1 convolutions to reduce computational complexity while maintaining performance.

#### 5.6.1.2 Model pruning

Techniques like parameter pruning, quantisation, and knowledge distillation are employed to compress pre-trained networks, making them more suitable for resource-limited devices.

For example, pruning removes unnecessary parameters, knowledge distillation moves information from a larger, more complicated model to a smaller, simpler one, while quantisation decreases the accuracy of weights.

### 5.6.2 Co-inference at the edge

When models are too large for direct deployment on devices, workload redistribution and splitting approaches are employed.

#### 5.6.2.1 *Workload redistribution*

Computation can be offloaded to edge nodes or cloud servers, balancing between increased computational capacity and the overhead of data transmission. Optimisation-based offloading strategies are designed to manage constraints like latency, bandwidth, and energy consumption.

#### 5.6.2.2 *Model splitting*

DNN models can be divided into segments, with certain layers executed on end devices and others offloaded to cloud servers or edge servers. This approach enhances performance by reducing latency and power consumption while leveraging the computation capabilities of distributed resources.

#### 5.6.2.3 *Model early termination*

This technique allows a DNN to complete inference early if a high confidence result is reached at an intermediate layer. This reduces the need for further computation and data transmission, optimising both latency and energy efficiency.

### 5.6.3 *Private inference*

Privacy concerns arise when data, potentially containing sensitive information, is transmitted to cloud or edge servers for inference. Two key techniques are used to enhance privacy.

#### 5.6.3.1 *Private computation*

Secure multiparty computation and homomorphic encryption are two techniques that allow encrypted data to be processed without disclosing its contents. These approaches ensure that data remains private while still allowing for effective AI inference.

#### 5.6.3.2 *Data anonymisation*

By adding noise to the data, anonymisation techniques protect sensitive information during inference. Approaches like differential privacy inject noise into data, ensuring that sensitive features are concealed while preserving the ability to perform accurate inference.

Overall, these enabling technologies facilitate the deployment of AI inference in resource-constrained environments, optimising the balance between latency, computational efficiency, accuracy, and privacy.

## 5.7 Collaborative computation an edge and cloud architecture

In edge and cloud computing, DNN processing can be optimised by distributing tasks between different devices and servers as explained in Figure 5.3 [4].

The three prominent key strategies are as follows.

### 5.7.1 *Partial offloading*

Some parts of the DNN are processed on the end device, while other parts are sent to a cloud or edge server for computation. This method helps reduce latency by sending smaller, processed information instead of large raw information over the network.

Tools like "Neurosurgeon" [46] help decide where to split the DNN tasks between the device and the server, while "DeepThings" focuses on splitting the data itself, which is useful for devices with limited memory.

### 5.7.2 *Hierarchical architectures*

Computation is divided between different layers, with simpler tasks handled by edge servers and more complex processing done in the cloud.

For example, in a Distributed Deep Neural Network [47] system, some layers of the DNN are processed on end devices or edge servers, while the rest is completed in the cloud. This setup allows quick, local processing and improves system reliability and privacy.

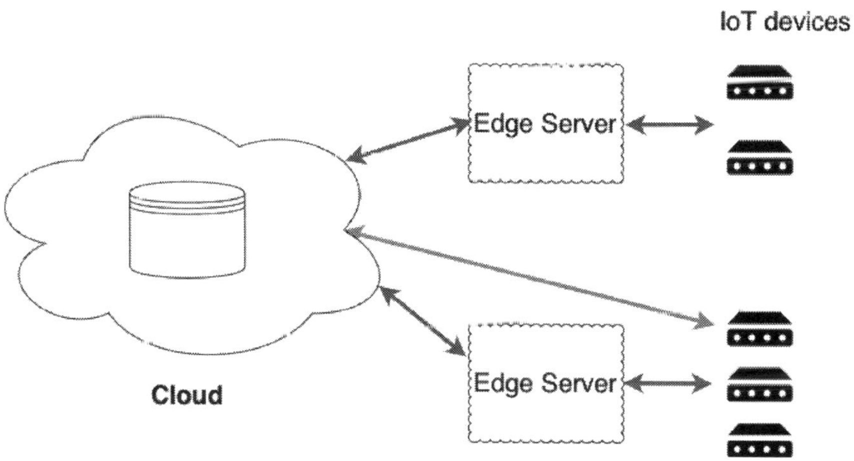


*Figure 5.3* Collaborative computation among devices and edge and cloud servers.

### 5.7.3 *Distributed computing*

Instead of relying on a single powerful server, DNN tasks are spread across multiple peer devices, like smartphones or small computers. The distribution of tasks depends on the computing power and memory of each device, allowing for a flexible and effective use of resources. These strategies aim to enhance the performance of DNNs by effectively using the combined power of edge devices, servers, and the cloud.

## 5.8 Privacy in edge computing and collaborative computation

When data are transferred between edge servers, the cloud, and end devices in edge computing or collaborative computation scenarios, it may contain sensitive information, raising privacy concerns. Given that edge servers operate within specific geographical areas, the source of the data can often be traced, making privacy protection crucial. While edge computing naturally reduces the amount of data shared over networks, thereby improving privacy, additional measures can further safeguard sensitive information.

### 5.8.1 *Adding noise to data*

A common strategy for preserving data privacy involves intentionally adding noise to the data before transmission [48]. On edge devices, key features can be extracted and perturbed with noise prior to being sent to the cloud. A lightweight deep neural network (DNN) on the edge performs initial processing, while a cloud-based DNN—trained to interpret noisy inputs—completes the classification task. This dual-stage processing approach leverages the principles of differential privacy to protect sensitive information while maintaining model performance.

### 5.8.2 *Cryptographic methods*

During processing, data can be secured using cryptographic techniques. "Homomorphic encryption" enables computations with encrypted data, so neither the end device nor the edge server can access the other's information.

For example:

(i) CryptoNets: In CryptoNets, DNN operations are transformed to work with encrypted data, though this process can be slow.
(ii) Secure Multiparty Computation: In secure multiparty computation, multiple devices work to compute a result while keeping their intermediate data private. Although effective for maintaining privacy, this technique can be communication concentrated.

These strategies help protect data privacy in edge computing environments, tackling the challenges of secure information processing and communication.

## 5.9 Open challenges and future directions

AIoT, the combination of the IoT and AI, is still developing and faces several challenges. The following are some key issues and future directions:

### *5.9.1 Cross-platform compatibility*

AIoT involves many different devices, like Raspberry Pi, FPGA boards, and smartphones, each using various sensors. These devices need to work together smoothly, even though they use different communication methods like Bluetooth or Wireless Fidelity (Wi-Fi). This diversity makes it hard to ensure all devices can interconnect and work together effectively.

To solve this, researchers are looking into new ways to manage these devices more easily, using technologies like Network Function Virtualisation and Software-Defined Networking. Network Function Virtualisation can make network functions more flexible and Software-Defined Networking can help by allowing easier control over different devices. Creating standard communication rules for all these devices could also help improve how they interact.

### *5.9.2 Resource management*

AIoT systems often use many sensors and devices that have limited electrical power and computing power. These devices can't always handle complex tasks on their own. To make the best use of available resources, tasks can be broken down and shared among different devices.

Because AIoT environments are constantly changing, it's important to manage resources in real time. New strategies are needed to balance computing, communication, and data storage effectively. Researchers are also using AI techniques, like Deep Reinforcement Learning, to help manage these resources.

### *5.9.3 Building and implementing AI models*

AI models need to be compressed and accelerated to work well on devices with limited resources. This often requires a lot of fine-tuning and can sometimes reduce the model's effectiveness.

To improve this, there is ongoing research into making these compression and acceleration techniques more automatic and adaptable. Additionally, training AI models can be tricky in a distributed environment. Federated

Learning helps by allowing several devices to train models together without sharing sensitive information. However, challenges like slow communication and bandwidth issues still need to be addressed. Better communication methods and training techniques are needed to improve efficiency.

### 5.9.4 *Data protection*

AIoT systems can be vulnerable to security threats and privacy issues. Devices at the edge of the network can be targeted by attacks like Distributed Denial of Service. These devices often have limited resources, making it hard to protect them effectively. Research is needed to develop lighter security solutions that use fewer resources.

Privacy is also a concern because AIoT devices collect and store sensitive data. To protect these data, techniques like Federated Learning, differential privacy, and encryption are being explored. Blockchain technology could also help, but it needs to be improved to use network resources more efficiently.

### 5.9.5 *Moral implications in AIoT*

AIoT systems make decisions quickly, often without human oversight. It's important to ensure these systems are fair, honest, accountable, safe, and sustainable. This means AI should not discriminate, should be transparent, have responsible developers, be accurate and secure, and minimise environmental impact. These ethical principles help guide the development and practice of AIoT technologies.

## 5.10 Conclusion

The integration of AI and the IoT, supported by edge computing, is poised to transform numerous industries by enabling real-time decision-making, enhancing data privacy, and reducing latency. While traditional cloud-centric architectures have supported the growth of AI and IoT, they face significant challenges in terms of scalability, privacy, and efficiency. The emerging paradigm of edge ML addresses these issues by decentralising computation, allowing data to be processed closer to its source.

Key techniques such as Federated Learning, Knowledge Distillation, and Transfer Learning have been highlighted as innovative solutions to overcome the constraints of edge devices, which are often limited in computational power and memory. By leveraging these techniques, AI models can be trained and deployed efficiently in resource-constrained environments, ensuring that the benefits of AIoT can be realised even at the network's edge.

Moreover, the concept of the AIoT underscores the potential for AI models to be tightly integrated with IoT systems, enabling more intelligent and autonomous operations across various sectors, including industrial automation, healthcare, and intelligent cities. However, challenges such as optimising AI models for edge deployment, managing distributed resources, and ensuring security remain critical areas for further research and development.

In summary, while the integration of AI and IoT presents a promising future, it also demands continuous innovation and adaptation to address the technical and practical challenges inherent in deploying intelligent systems at the edge. The ongoing advancements in edge computing and AI methodologies are crucial to fully unlocking the potential of AIoT, paving the way for smarter, more resourceful, and more secure applications across diverse industries.

## References

[1] Khouas, Aymen Rayane, Mohamed Reda Bouadjenek, Hakim Hacid, and Sunil Aryal. "Training Machine Learning Models at the Edge: A Survey." *arXiv preprint arXiv:2403.02619* (2024).

[2] Hua, Haochen, Yutong Li, Tonghe Wang, Nanqing Dong, Wei Li, and Junwei Cao. "Edge computing with artificial intelligence: A machine learning perspective." *ACM Computing Surveys* 55, no. 9 (2023): 1–35.

[3] Chang, Zhuoqing, Shubo Liu, Xingxing Xiong, Zhaohui Cai, and Guoqing Tu. "A survey of recent advances in edge-computing-powered artificial intelligence of things." *IEEE Internet of Things Journal* 8, no. 18 (2021): 13849–13875.

[4] Merenda, Massimo, Carlo Porcaro, and Demetrio Iero. "Edge machine learning for ai-enabled IoT devices: A review." *Sensors* 20, no. 9 (2020): 2533.

[5] Ghosh, Ashish, Debasrita Chakraborty, and Anwesha Law. "Artificial intelligence in Internet of things." *CAAI Transactions on Intelligence Technology* 3, no. 4 (2018): 208–218.

[6] Din, Ikram Ud, Mohsen Guizani, Joel J. P. C. Rodrigues, Suhaidi Hassan, and Valery V. Korotaev. "Machine learning in the Internet of Things: Designed techniques for smart cities." *Future Generation Computer Systems* 100 (2019): 826–843.

[7] Deng, Shuiguang, Hailiang Zhao, Weijia Fang, Jianwei Yin, Schahram Dustdar, and Albert Y. Zomaya. "Edge intelligence: The confluence of edge computing and artificial intelligence." *IEEE Internet of Things Journal* 7, no. 8 (2020): 7457–7469.

[8] Cui, Laizhong, Shu Yang, Fei Chen, Zhong Ming, Nan Lu, and Jing Qin. "A survey on application of machine learning for Internet of Things." *International Journal of Machine Learning and Cybernetics* 9 (2018): 1399–1417.

[9] Mohammadi, Mehdi, Ala Al-Fuqaha, Sameh Sorour, and Mohsen Guizani. "Deep learning for IoT big data and streaming analytics: A survey." *IEEE Communications Surveys & Tutorials* 20, no. 4 (2018): 2923–2960.

[10] Chen, Yanjiao, Baolin Zheng, Zihan Zhang, Qian Wang, Chao Shen, and Qian Zhang. "Deep learning on mobile and embedded devices: State-of-the-art, challenges, and future directions." *ACM Computing Surveys (CSUR)* 53, no. 4 (2020): 1–37.

[11] Li, He, Kaoru Ota, and Mianxiong Dong. "Learning IoT in edge: Deep learning for the Internet of Things with edge computing." *IEEE Network* 32, no. 1 (2018): 96–101.

[12] Sodhro, Ali Hassan, Sandeep Pirbhulal, and Victor Hugo C. De Albuquerque. "Artificial intelligence-driven mechanism for edge computing-based industrial

applications." *IEEE Transactions on Industrial Informatics* 15, no. 7 (2019): 4235–4243.
[13] Dai, Yueyue, Du Xu, Sabita Maharjan, Guanhua Qiao, and Yan Zhang. "Artificial intelligence empowered edge computing and caching for internet of vehicles." *IEEE Wireless Communications* 26, no. 3 (2019): 12–18.
[14] Ji, Hongjing, Osama Alfarraj, and Amr Tolba. "Artificial intelligence-empowered edge of vehicles: Architecture, enabling technologies, and applications." *IEEE Access* 8 (2020): 61020–61034.
[15] McMahan, Brendan, Eider Moore, Daniel Ramage, Seth Hampson, and Blaise Aguera y Arcas. "Communication-efficient learning of deep networks from decentralized data." In *Artificial Intelligence and Statistics*, pp. 1273–1282. PMLR, 2017.
[16] Ye, Yunfan, Shen Li, Fang Liu, Yonghao Tang, and Wanting Hu. "EdgeFed: Optimized federated learning based on edge computing." *IEEE Access* 8 (2020): 209191–209198.
[17] Liu, Ruixuan, Yang Cao, Masatoshi Yoshikawa, and Hong Chen. "Fedsel: Federated SGD under local differential privacy with top-k dimension selection." In *Database Systems for Advanced Applications: 25th International Conference, DASFAA 2020, Jeju, South Korea, September 24–27, 2020, Proceedings, Part I 25*, pp. 485–501. Springer International Publishing, 2020.
[18] Yang, Kai, Tao Jiang, Yuanming Shi, and Zhi Ding. "Federated learning based on over-the-air computation." In *ICC 2019–2019 IEEE International Conference on Communications (ICC)*, pp. 1–6. IEEE, 2019.
[19] Thapa, Chandra, Pathum Chamikara Mahawaga Arachchige, Seyit Camtepe, and Lichao Sun. "Splitfed: When federated learning meets split learning." *Proceedings of the AAAI Conference on Artificial Intelligence*, vol. 36, no. 8 (2022): 8485–8493.
[20] Wang, Yue, Zhi Tian, Xin Fan, Yan Huo, Cameron Nowzari, and Kai Zeng. "Distributed swarm learning for internet of things at the edge: Where artificial intelligence meets biological intelligence." *arXiv preprint arXiv:2210.16705* (2022).
[21] Hegedűs, István, Gábor Danner, and Márk Jelasity. "Gossip learning as a decentralized alternative to federated learning." In *Distributed Applications and Interoperable Systems: 19th IFIP WG 6.1 International Conference, DAIS 2019, Held as Part of the 14th International Federated Conference on Distributed Computing Techniques, DisCoTec 2019, Kongens Lyngby, Denmark, June 17–21, 2019, Proceedings 19*, pp. 74–90. Springer International Publishing, 2019.
[22] Pan, Sinno Jialin, and Qiang Yang. "A survey on transfer learning." *IEEE Transactions on Knowledge and Data Engineering* 22, no. 10 (2009): 1345–1359.
[23] Wu, Yue, Yinpeng Chen, Lijuan Wang, Yuancheng Ye, Zicheng Liu, Yandong Guo, and Yun Fu. "Large scale incremental learning." In *Proceedings of the IEEE/CVF Conference on Computer Vision and Pattern Recognition*, pp. 374–382. IEEE Xplore, 2019.
[24] Nichol, Alex, Joshua Achiam, and John Schulman. "On first-order meta-learning algorithms." *arXiv preprint arXiv:1803.02999* (2018).
[25] Gou, Jianping, Baosheng Yu, Stephen J. Maybank, and Dacheng Tao. "Knowledge distillation: A survey." *International Journal of Computer Vision* 129, no. 6 (2021): 1789–1819.
[26] Gholami, Amir, Sehoon Kim, Zhen Dong, Zhewei Yao, Michael W. Mahoney, and Kurt Keutzer. "A survey of quantization methods for efficient neural network inference." In *Low-Power Computer Vision*, pp. 291–326. Chapman and Hall/CRC, 2022.

[27] Zhou, Xinlei, and Dasen Yan. "Model tree pruning." *International Journal of Machine Learning and Cybernetics* 10 (2019): 3431–3444.
[28] Guo, Yan, Shangguang Wang, Ao Zhou, Jinliang Xu, Jie Yuan, and Ching-Hsien Hsu. "User allocation-aware edge cloud placement in mobile edge computing." *Software: Practice and Experience* 50, no. 5 (2020): 489–502.
[29] Kala, Rahul. *Autonomous Mobile Robots Planning, Navigation and Simulation*. Academic Press, 2023.
[30] Wang, Yixuan, Kun Wang, Huawei Huang, Toshiaki Miyazaki, and Song Guo. "Traffic and computation co-offloading with reinforcement learning in fog computing for industrial applications." *IEEE Transactions on Industrial Informatics* 15, no. 2 (2018): 976–986.
[31] Courbariaux, Matthieu, Itay Hubara, Daniel Soudry, Ran El-Yaniv, and Yoshua Bengio. "Binarized neural networks: Training deep neural networks with weights and activations constrained to+ 1 or−1." *arXiv preprint arXiv:1602.02830* (2016).
[32] Gerstner, Wulfram, and Werner M. Kistler. *Spiking Neuron Models: Single Neurons, Populations, Plasticity*. Cambridge University Press, 2002.
[33] Hinton, Geoffrey. "The forward-forward algorithm: Some preliminary investigations." *arXiv preprint arXiv:2212.13345* (2022).
[34] Goodfellow, Ian, Jean Pouget-Abadie, Mehdi Mirza, Bing Xu, David Warde-Farley, Sherjil Ozair, Aaron Courville, and Yoshua Bengio. "Generative adversarial nets." *Advances in Neural Information Processing Systems* 27 (2014).
[35] Altman, Naomi S. "An introduction to kernel and nearest-neighbor nonparametric regression." *The American Statistician* 46, no. 3 (1992): 175–185.
[36] Gope, Dibakar, Ganesh Dasika, and Matthew Mattina. "Ternary hybrid neural-tree networks for highly constrained IoT applications." *Proceedings of Machine Learning and Systems* 1 (2019): 190–200.
[37] Chen, Chen, Bin Liu, Shaohua Wan, Peng Qiao, and Qingqi Pei. "An edge traffic flow detection scheme based on deep learning in an intelligent transportation system." *IEEE Transactions on Intelligent Transportation Systems* 22, no. 3 (2020): 1840–1852.
[38] Liu, Yi, Chao Yang, Li Jiang, Shengli Xie, and Yan Zhang. "Intelligent edge computing for IoT-based energy management in smart cities." *IEEE Network* 33, no. 2 (2019): 111–117.
[39] Hu, Long, Yiming Miao, Gaoxiang Wu, Mohammad Mehedi Hassan, and Iztok Humar. "iRobot-factory: An intelligent robot factory based on cognitive manufacturing and edge computing." *Future Generation Computer Systems* 90 (2019): 569–577.
[40] Ning, Zhaolong, Peiran Dong, Xiaojie Wang, Lei Guo, and Joel JPC Rodrigues. "Xiangjie Kong, Jun Huang, and Ricky YK Kwok. Deep reinforcement learning for intelligent internet of vehicles: An energy efficient computational offloading scheme." *IEEE Transactions on Cognitive Communications and Networking* 5, no. 4 (2019): 1060–1072.
[41] Kala, Rahul. *On-Road Intelligent Vehicles Motion Planning for Intelligent Transportation Systems*. Elsevier, 2016.
[42] Chollet, François. "Xception: Deep learning with depthwise separable convolutions." In *Proceedings of the IEEE Conference on Computer Vision and Pattern Recognition*, pp. 1251–1258. IEEE, 2017.
[43] Ding, Wei, Zeyu Huang, Zunkai Huang, Li Tian, Hui Wang, and Songlin Feng. "Designing efficient accelerator of depthwise separable convolutional neural network on FPGA." *Journal of Systems Architecture* 97 (2019): 278–286.

[44] Redmon, Joseph, Santosh Divvala, Ross Girshick, and Ali Farhadi. "You only look once: Unified, real-time object detection." In *Proceedings of the IEEE Conference on Computer Vision and Pattern Recognition*, pp. 779–788. IEEE, 2016.
[45] Iandola, Forrest N., et al. "SqueezeNet: AlexNet-level accuracy with 50x fewer parameters and< 0.5 MB model size." *arXiv preprint arXiv:1602.07360* (2016).
[46] Kang, Yiping, Johann Hauswald, Cao Gao, Austin Rovinski, Trevor Mudge, Jason Mars, and Lingjia Tang. "Neurosurgeon: Collaborative intelligence between the cloud and mobile edge." *ACM SIGARCH Computer Architecture News* 45, no. 1 (2017): 615–629.
[47] Teerapittayanon, Surat, Bradley McDanel, and Hsiang-Tsung Kung. "Distributed deep neural networks over the cloud, the edge and end devices." In *2017 IEEE 37th International Conference on Distributed Computing Systems (ICDCS)*, pp. 328–339. IEEE, 2017.
[48] Wang, Ji, Jianguo Zhang, Weidong Bao, Xiaomin Zhu, Bokai Cao, and Philip S. Yu. "Not just privacy: Improving performance of private deep learning in mobile cloud." In *Proceedings of the 24th ACM SIGKDD International Conference on Knowledge Discovery & Data Mining*, pp. 2407–2416. ACM, 2018.

Chapter 6

# Blockchain-enabled edge intelligence for IoT

*Gauri Shankar, Md Raihan Uddin, and Shareeful Islam*

## 6.1 Introduction

### *6.1.1 Edge intelligence and IoT*

The growing number of devices and sensors in the Internet of Things (IoT) networks generate data on a large scale. In the traditional and centralised IoT infrastructure, the data collected from the edge devices, including sensor data and mobile devices, are sent to the cloud or data processing centres. These data are analysed and used for training and computation to enhance the performance of IoT applications and the connected devices on the IoT network. In recent years, the introduction of the fifth-generation (5G) network communication technology has expressed more potential in the IoT infrastructure, which has increased IoT adoption on a large scale for applications such as healthcare, smart cities, and industry [1]. Adopting this advanced communication technology develops a network of devices that generates data in different formats, such as text, audio, images, and videos. This development in IoT introduces more advanced and complex IoT applications that create substantial amounts of data in a short time on the network [2]. However, there is a significant improvement in the quality of data collected by edge devices and in communication with the central server. But the, serious issues arise with the data processing analysis, as the central server still performs the computation of the collected data. In such cases, the latency of the transmitting data to the server is improved but is lacking in real-time data processing and decision-making. Because the size of the files is large and the speed at which they are collected on the central server is not fast enough, in 2010, significant advancements in Deep Learning (DL) sparked widespread interest in artificial intelligence (AI) for IoT networks, a potential technology to address these issues in the IoT [3].

Integrating AI on the edge devices of the IoT introduced a modern technology known as Edge Intelligence (EI). This combines the power of edge computing and AI to process and analyse the data collected from the sensors on the edge devices. EI satisfies the critical requirements of the IoT, such as

DOI: 10.1201/9781003512066-6

real-time data processing and decision-making, solves the issue of IoT related to centralised processing and reduces the latency between data generation and actionable insights, improves security and privacy by keeping sensitive data closer to its source, reducing the risk of data breaches and enhancing privacy protection. EI provides scalable data processing capabilities and storage for the increasing volume of data generated by IoT devices. Scalability and distributed processing capabilities allow the system to scale efficiently as the number of IoT devices increases. The decentralised process provides reliability on the network by ensuring continuous data processing, even if some edge nodes stop data processing [4]. The benefits of EI on IoT are shown in Figure 6.1.

The EI and advanced communication technology such as 5G are integrating with IoT, advancing the critical role of the IoT for enterprises and consumers. As a result, the adoption of IoT networks and spending on the IoT infrastructure significantly increased, as represented in Figures 6.2 and 6.3. In Figure 6.2, the global spending cases on IoT are represented as, in 2018, the spending was $646 billion, which is increasing, and in 2023, the spending reached $805.7 billion. However, there is a slight deflection in the year 2022 because the global pandemic affected the manufacturing of IoT devices [5].

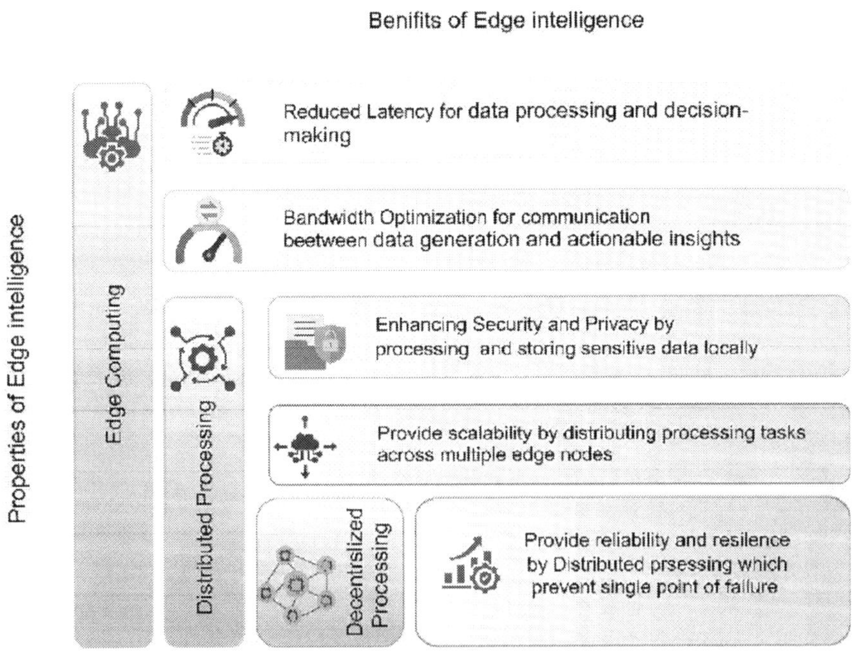


*Figure 6.1* Properties of Edge Intelligence and its benefits.

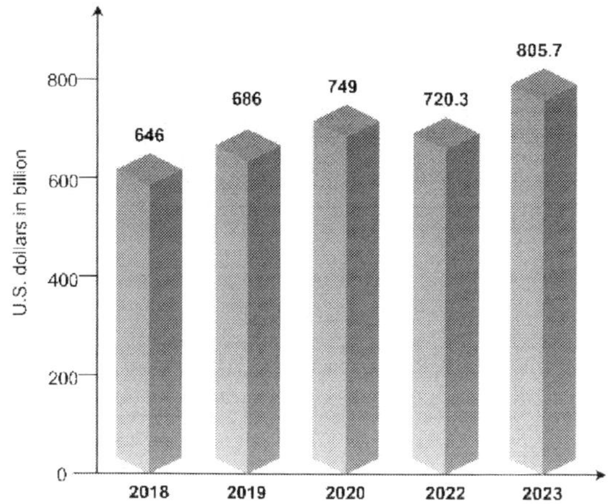


*Figure 6.2* Worldwide spending on IoT from 2018 to 2023 [5].

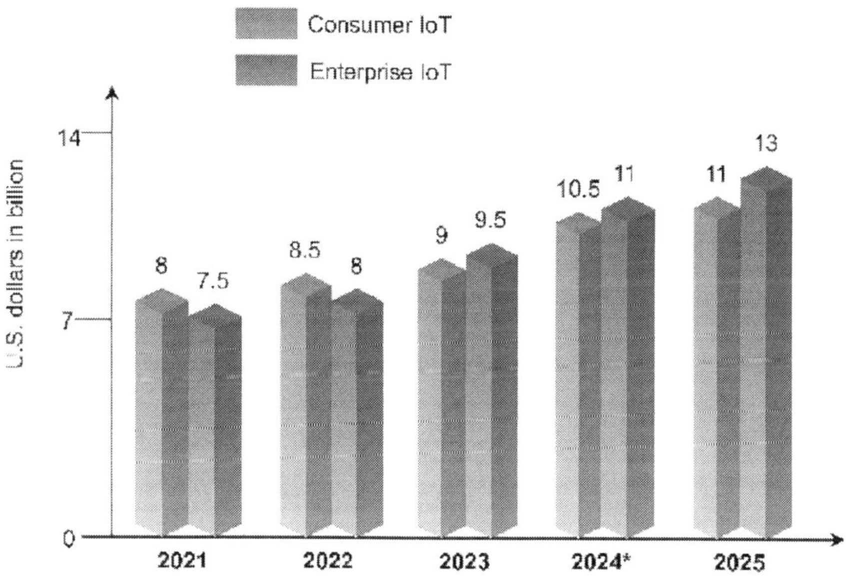


*Figure 6.3* Worldwide IoT connection from 2021 to 2025 by application [6].

Figure 6.3 shows the number of worldwide connected IoT devices for consumers and enterprises through the application. In 2021, the total number of connected IoT devices was 15.5 billion in the world, which will increase in 2023 by 18.5 billion. In the future, it is predicted to reach around 24 billion by the year 2025 [6].

### 6.1.2 *Blockchain in brief*

The term blockchain began getting recognition in 2008 when a white paper written by a pseudonymous person titled 'Bitcoin: A Peer-to-Peer Electronic Cash System' was published [7]. The basic architecture of bitcoin involves the Markle tree, Hash cash proof system, and public key cryptography to develop a decentralised and distributed network of a cryptocurrency that runs on the proof of work consensus algorithm. Before the introduction of bitcoin, the transaction went through a centralised third-party server/broker in the financial trading system, which is highly vulnerable to corruption or a single point of failure. At the same time, the decentralised and distributed property of the bitcoins remove these issues. In the present scenarios in digital transformation, the adoption of blockchain is growing very rapidly because of its properties, which can provide security and privacy and fault-tolerant systems through advanced asymmetric encryption algorithms and consensus mechanisms. The decentralised and distributed property of the blockchain offers peer-to-peer (P2P) transactions without any third-party interference, which motivates the current edge computing devices to adopt blockchain. This will remove centralised cloud computing from the IoT architecture [8].

The blockchain is an immutable ledger system that provides a decentralised record of transactions executing on the network. This list of transactions is stored in the network's encrypted form and develops a transaction block. The network nodes run on the consensus algorithm to add transactions to the blockchain block. The consensus is designed to verify the authenticity of the transaction, validate the transaction, and add to the block based on the outcome. All the blocks created in the blockchain are connected immutably using the Merkle tree [9]. The features of the blockchain influence the adoption of other technologies, such as IoT, by ensuring the security and privacy of the data, cost-efficiency, authenticity, and auditability. In the past few years, the adoption of blockchain has increased the market size of IoT infrastructure. As of 2020, the global market size of the blockchain for IoT was $258 million, which was exposed to an increase of 45% of the compound annual growth rate and expected to reach up to $2409 million by 2026 [10].

### 6.1.3 *Contribution of the chapter*

Although EI is advancing with the IoT technology with advances of 5G communication and artificial intelligence (AI) in edge computing, it is still lacking due to specific challenges related to security and privacy, power consumption cost, data management, and regulations compliance in the system. Based on these challenges and previous studies, we developed the following research questions:

**RQ1:** What factors of blockchain-enabled edge intelligence can solve the issues of high-power consumption in intelligent IoT systems?

**RQ2:** At what level does blockchain help to protect data security, user privacy, and vulnerabilities on edge devices?

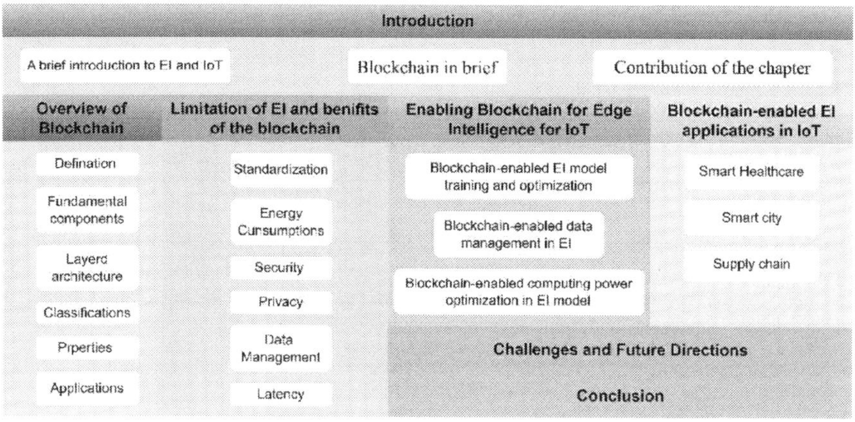


*Figure 6.4* The structure of this chapter.

**RQ3:** What challenges and factors will occur in integrating the blockchain with edge intelligence?

Furthermore, the structure of this chapter is as depicted in Figure 6.4.

## 6.2 Overview of blockchain

### 6.2.1 *Blockchain definition*

Blockchain is a distributed ledger accessed by all the connected devices on a network in P2P decentralised communication. This ensures transparency, trust, immutability, and availability of stored transaction records. In the ledger, all connected devices can write the transaction with an anonymous identity, as it provides secure cryptographic encryption to identity and transaction [9]. In short, 'Blockchain is a cryptographically secure distributed ledger technology that ensures anonymity while writing transactions on the ledger by participants that connected in a peer-to-peer decentralised network and stored an immutable copy of the ledger'.

### 6.2.2 *Fundamental components of the blockchain*

The development of any blockchain network requires the following fundamental components.

#### *6.2.2.1 P2P decentralised network*

A P2P decentralised network offers direct communication between devices without any centralised medium to endorse the communication process. In this communication model, a single device can be used as a client while requesting data from another device or as a server while responding to other devices' requests for files or data. In the blockchain, this communication model improves communication and network stability by ensuring faster and unbiased services to each connected device compared to centralised communication. This communication model provides a trustless medium for communication where devices do not need to be familiar with each other to share the data; each connected node will be notified about the modification in the network. If the modification in the network communication is unverified, then the communication process will be rejected based on the vote performed by each connected node by validating the communication. This model removes the concept of a single point of failure and prevents corruption with fault-tolerant distributed resources [11].

#### *6.2.2.2 Distributed ledger*

The distributed ledger is a database available to all connected devices in the P2P decentralised network. The copy of this database is stored on all connected devices and updated synchronously when any node does any modification. As the update on the database can be seen by all the connected devices, it provides visibility and transparency in the network and improves trust. This distributed ledger removes the centralised system on the network's database, making it resilient to corruption and cyberattacks and providing prevention from a single point of failure. Combining a distributed ledger and a decentralised network enhances the distributed network and ensures only legitimate database modification through voting and validating communication. If any corrupted device is found, it can be removed from the network without affecting the other devices. Due to the absence of centralised processing, the communication speed is fast, and the cost is low. It also provides fault tolerance, as the database is available on all the connected devices, so the unavailability of any device does not affect the network's communication. In the case of blockchain networks, the database stores the hash of the internet protocol (IP) address of devices and shared data between them, making it a more secure and privacy-oriented technology [12].

#### *6.2.2.3 Cryptographic protocols*

The cryptographic protocol ensures data security and the privacy of the blockchain users. The cryptographic protocol focuses on the confidentiality, integrity, and availability (CIA) triad of cybersecurity. The CIA triad ensures the data's security and privacy while it is available to the users. Cryptography is converting plain text into cypher text, which is done by the encryption algorithm, and converting back to plain text from cypher text is performed with the decryption algorithm. Several cryptographic protocols are integrated into the blockchain to achieve high security and privacy, such as Cryptographic Hash Functions (CHF), Public key Cryptography (PKC), and digital signature [13].

*The CHF function:* It is performed on any arbitrary size of data to generate a fixed-length alphanumeric stream of characters known as a hash, and in the blockchain, the 256-bit hash is used. The process of hashing is a one-way function, meaning getting the arbitrary size of data back from the hash is impossible. The properties of the cryptographic hash functions make it unique. One is collision resistance, which works behind the concept of uniqueness in the hash, as two different data never have the same hash value. Let *N* and *M* be two arbitrary data; H is the hash function, so *H(N)! = H(M)*. However, there are some hashing algorithms that contradict these properties. The second property of the CHF is Hiding, which is a developed cause of a one-way function in which the original message can never be retrieved from the hash. *N* can never be calculated from H*(S||N)*, where *S* is a pseudorandom number. The last and main property is puzzle-friendliness of the CHF, which says that in the n-bit output of the *M*, if *k* is selected as high mean entropy, then *H(k||N) = Min in time by* $2^n$. Markel tree is used to compute the hash and chained with the hash of the previous block, message digest and cryptographic wallet hash function [13].

*The immutability:* In the blockchain, immutability is ensured by the CHF. When the transaction is added to the block of the blockchain, the computation of consensus algorithms gets more complex, which strengthens the immutability of the blockchain so that the latest block will be more immutable than the first block. This immutability is achieved by hashing the information in the previous block of the transaction, adding that hash in the latest block, generating a hash for the current block, and so on. By this process, a tree of hashes is developed known as a Markle Tree. If any of the previous data are modified, all the hashes are changed up to the current block, making the chain invalid. Figure 6.5 presents the structure of the Markle tree [14].

Public Key Cryptography (PKC) is an asymmetric version of the cryptographic protocol that uses two different pseudorandom strings with encryption and

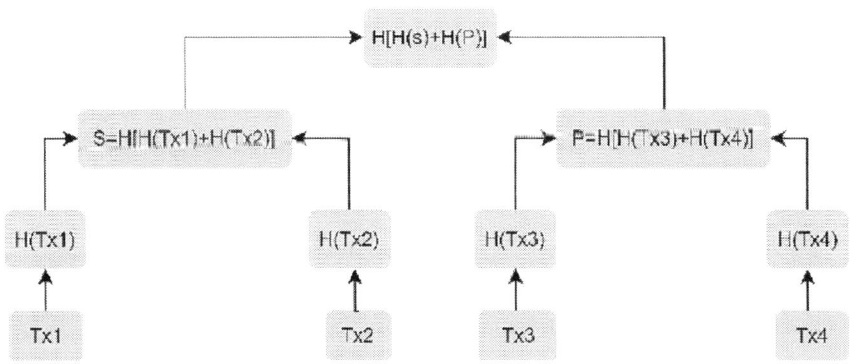


*Figure 6.5* Markle tree in blockchain.

decryption algorithms to secure data while transmitting over the blockchain. The string used for the encryption of the information is known as the public key for a receiver used by the sender, and the key used in the receiver's decryption is the receiver's private key. The public key is publicly available in the blockchain network as a public ID. In public key cryptography, if $K_1$ is the secret key and $K_2$ is the public key, then $K_1 != K2$. Some of the most used PKCs in the blockchain include Rivest–Shamir–Adleman (RSA), Diffie–Hellman, ElGamal, and Elliptic Curve Cryptography (ECC). The PKC provides security and privacy to the user's information by allowing data access only to the authorised user. The PKC is also used as an access credential for the blockchain network, with a public key as a username, and a private key as a password. The PKC process can be better understood from Figure 6.6 [15].

The PKC in the blockchain consists of the following steps:

1) Key generation: Asymmetric cryptographic algorithms, such as RSA and ECC, are used to generate pairs of keys.
2) Encryption: The sender can use the receiver's public key to encrypt the data with the encryption algorithm derived from the cryptographic algorithm used to generate keys and transact the encrypted data into the distributed ledger.
3) Decryption: The receiver uses its private key with the decryption algorithm to decrypt the data from the ledger and can access the data upon successful decryption.

*Digital Signature:* The cryptographic protocol provides the authenticity of the communication in the blockchain. The digital signature process is the opposite of the encryption and decryption process, such that the message is signed by the sender's private key and verified by the sender's public ID (digital certificate/public key). The digital signature provides integrity, authentication, and non-repudiation in communication in the blockchain network. In the blockchain, the transaction is signed. In this, the sender sends the transaction on the network, which is hashed with

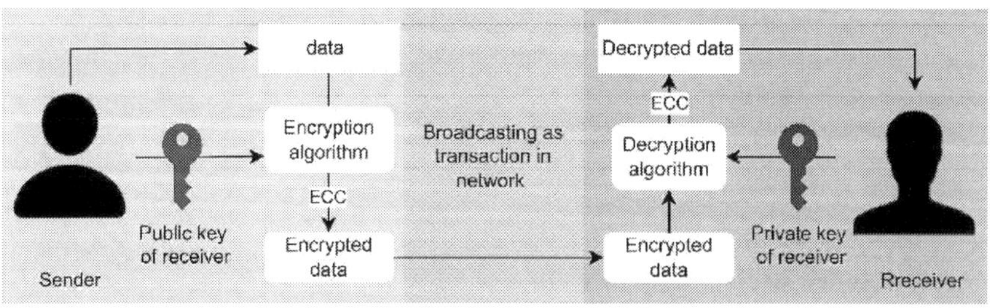


*Figure 6.6* Process of PKC in blockchain.

the Secure Hashing Algorithm (SHA 256-bits) and signed with its private key to generate the signature using signature generation algorithms. Later, the receiver verifies the authenticity of the transaction using the signer's public key, which is broadcasted as the signer's public ID on the network [13]. The process of the digital signature is represented in Figure 6.7.

The digital signature in the blockchain consists of the following steps:

1) Key generation: This can be done by public key cryptographic algorithms with digital signature variants such as Elliptic Curve Digital Signature Algorithm (ECDSA), Edward Curve Digital Signature Algorithm (EdDSA), and RSA.
2) Hashing: The communication is written on the network as a transaction, and it generates its hash using the SHA 256 Hashing algorithm, which is broadcast on the distributed ledger.
3) Signing a message: The transaction hash is signed using the receiver's public key (publicly available on the network as the Receiver's ID) on the signature generation algorithms mentioned in the first step and broadcast in the network.
4) Verification of the signature: At the receiver's side, the verification is performed using the receiver's private key in a verification algorithm and then matching the transaction hash from the network.

The identity of the signer remains anonymous throughout the digital signature process. The most popular available digital signatures are RSA, EdDSA, ECDSA Schnorr signature, and ElGamal signature. ECDSA and EdDCA are also broadly used in the blockchain.

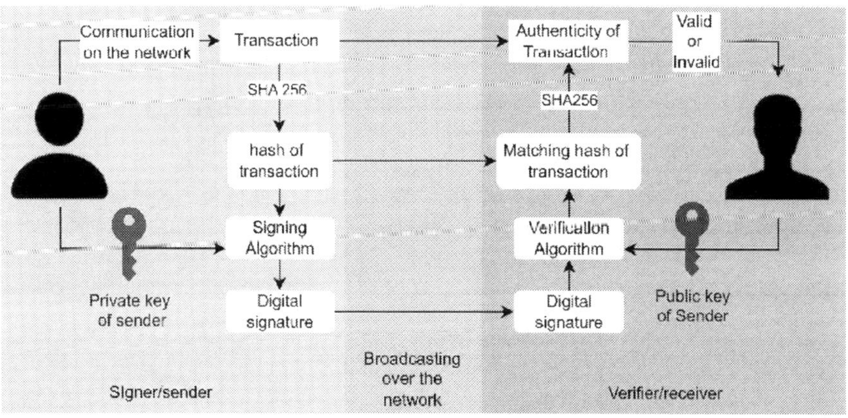


*Figure 6.7* Process of Digital signature in blockchain.

#### 6.2.2.4 *Consensus mechanism*

A fault-tolerant system in the blockchain is needed to achieve a common agreement for validating and verifying the transactions on the blockchain. This allows all the connected participants to have an equal chance to propose the agreement in support of acceptance of the rejection of the transaction. It helps to maintain the security and integrity of the blockchain network by verifying the validation of the transactions before adding to the blockchain. Besides this, it also helps the scalability of the blockchain with the growing number of devices [16]. Figure 6.8 represents a simple example of the consensus mechanism. Based on the work, four types of consensus mechanisms are identified as follows:

1) Resource-based: This type of consensus mechanism uses the computation power, space availability, and memory of the connected systems to validate and verify the transactions.
2) Economy-based: This allows the connected device with high wealth to compute and speed up the process of validation and verification of the transactions.
3) Capability-based: This is used to handle asset- and ownership-based transactions in the blockchain using a trusted execution environment (TEE).
4) Hybrid-based: This uses a combination of the resource-based, economy-based, and capability-based consensus mechanism characteristics.

In the blockchain, the following consensus are commonly used as listed in Table 6.1:

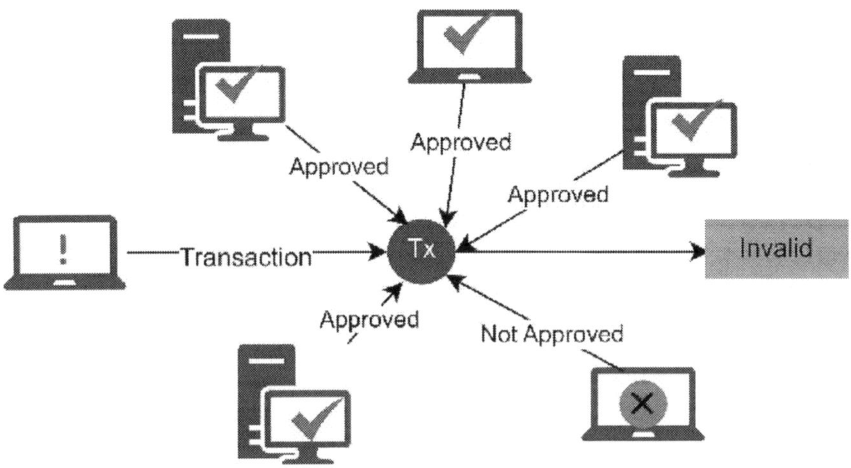


*Figure 6.8* Consensus mechanism in blockchain.

*Table 6.1* Common consensus mechanism

| *Consensus mechanism* | *Type* |
|---|---|
| Proof of Work (PoW) | Resource-based |
| Proof of Stake (PoS) | Economy-based |
| Delegated proof of stake (DPoS) | |
| Practical Byzantine Fault Tolerance (PBFT) | Capability-based |
| Proof of Authority (PoA) | |
| Proof of Elapsed Time (PET) | |

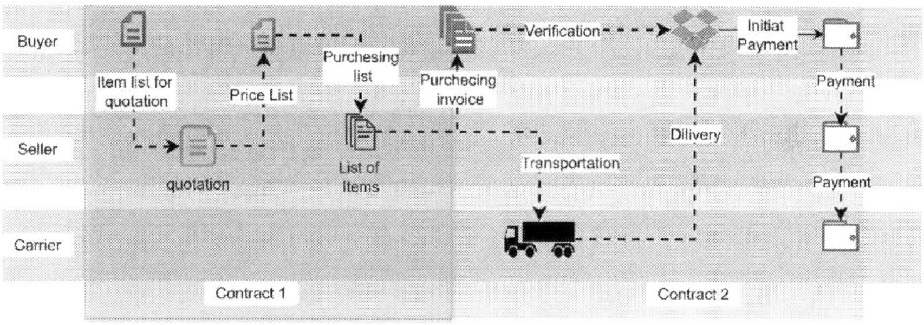


*Figure 6.9* Smart contract (An overview of smart contracts: challenges, advances, and platforms).

### *6.2.2.5 Smart contract*

A smart contract is a set of digital agreements that includes the protocol that helps to process these agreements between the parties. In technical terms, 'a smart contract is a self-execution set of instructions in the computer that execute only when it meets the predefined conditions'. The smart contract provides automation in the blockchain for the connected nodes, which helps to speed up the process in the blockchain [17]. The traditional contract required third parties. As the blockchain is decentralised and works on a P2P network, the case smart contract is directly executed between the agreeing parties without requiring a third party. The smart contract process is shown in Figure 6.9.

## ***6.2.3 Architecture of the blockchain***

The blockchain architecture can be divided into different layers based on the functionality and the components [18]. In the blockchain, the main layers are as presented in Figure 6.10.

*Physical Layer:* In the blockchain, the nodes are connected in a P2P communication for writing transactions, mining, and validation. These nodes

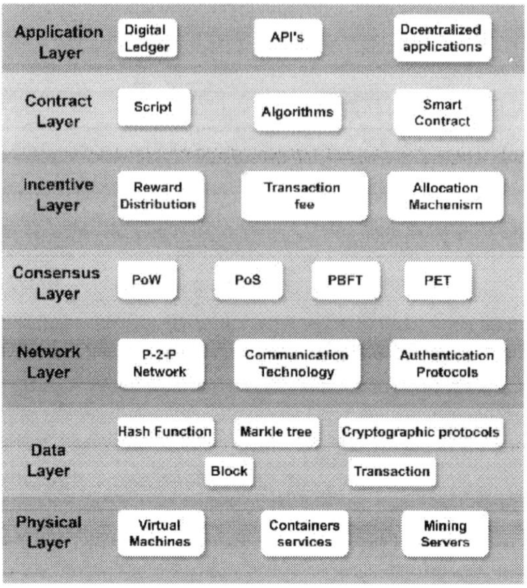


*Figure 6.10* Layered architecture of the blockchain.

are powerful computing devices that help to run the blockchain efficiently. Virtual machines such as Ethereum Virtual Machines (EVM) help run the blockchain framework on the computing devices. At the same time, mining servers validate and add transactions on the block, and the functionalities of these nodes are maintained with the container services.

*Data Layer:* This layer handles the records on the blockchain and encompasses the block, transaction, the hash functions that link blocks of transaction, the Markle tree, and the encryption algorithms to protect transaction communication. First, decryption is operated on the transaction, then the hash is matched with the previous block, added to the Markle tree, and finally written on the block.

*Network Layer:* All the nodes in the blockchain are connected to the decentralised P2P network using a communication medium. Because of decentralisation, each node acknowledges updates on the transactions on the ledger. Then, nodes can verify the transaction using an authentication protocol (digital signature). After the successful authentication, the transaction is sent to the next layer.

*Consensus layer:* This layer is responsible for the endorsement of the transaction based on the common roles agreed upon by every node. It also states that every node has an updated version of the ledger for the decentralised network. This helps in the governance of the blockchain in a decentralised environment and handles the generation and validation of the blocks.

*Incentive Layer:* The nodes collectively validate the transaction and add it to the blockchain. For this process, nodes use their computing power or resources, and based on the winner or used computing resource, some incentive is given to them as a reward. This incentive is generated from the transaction fee, which depends on the amount of the transaction. This is allocated to nodes using allocation mechanisms and is only used in public blockchains like Bitcoin and Ethereum.

*Contract layer:* This layer encapsulates the script and algorithms into a smart contract that helps to self-execute the predefined agreement for the parties participating in the transaction. The contract is authenticated by the parties using a digital signature, and it is broadcast in the network after successful execution using EVM.

*Application Layer:* This is the top layer of the blockchain, which is responsible for running Application Programming Interfaces (API) to provide access and the ability to program the blockchain. This layer defines the functionalities of the digital ledger, smart contract, and other applications.

### 6.2.4 *Classification of the blockchain*

Based on the access mechanism and functionalities [18], the blockchain is classified into three main categories, represented in Figure 6.11, which provides further explanation about each type of blockchain.

*Public Blockchain:* The public blockchain, or permissionless blockchain, is open to all, and any node can join the network anonymously. The public blockchain is fully decentralised and distributed, and each node has an equal chance of participating in the consensus mechanism. It is worked on based on a trustless system without involving a third party. Examples include Bitcoin and Ethereum.

*Private Blockchain:* The private or permissioned blockchain is an organisation's internal blockchain in which a central authority can manage nodes.

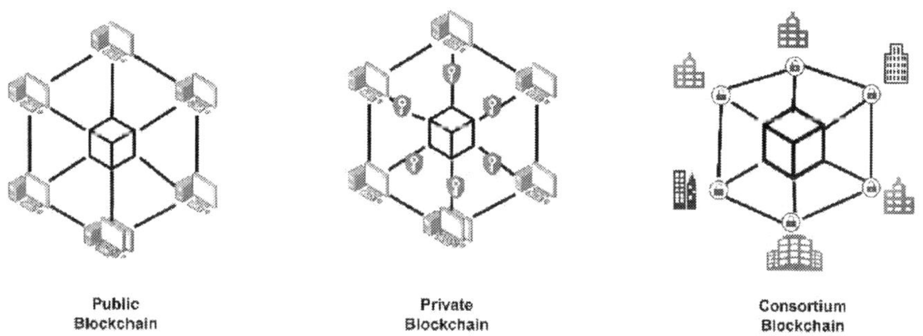


*Figure 6.11* Classification of the blockchain.

To join the private blockchain, a node must register with the organisation. In this type of blockchain, nodes operate based on defined policies and access control by the organisation. All nodes do not have equal chances to participate in consensus. Examples include Hyperledger and Corda.

*Consortium Blockchain:* This blockchain is also a permissioned blockchain. Participating organisations develop it. To join this blockchain, an organisation must agree to the rules of the consortium. Each organisation has its own blockchain and consensus mechanism in this blockchain, while the consortium runs its own consensus algorithm. An example is Hyperledger Fabric.

*Hybrid Blockchain:* A hybrid blockchain combines public and private blockchains, in which some nodes are permissionless and some are permissioned.

### 6.2.5 Properties of the blockchain

*Decentralisation:* The blockchain is not dependent on any centralised authority, which means it is free from a single point of failure. It provides trustless P2P communication for faster information sharing [18].

*Transparency:* The blockchain is transparent to every connected node. In the blockchain, every node maintains a synchronised copy of the distributed ledger locally. This provides transparency on the transaction ledger. Everyone except the receiver can see the transaction information but not the actual data inside the transaction.

*Scalability:* As the blockchain is decentralised and distributed, the number of nodes can increase without affecting the efficiency of the network. There are no centralised server capacity limits for storage and computing power, which makes this blockchain more scalable and reliable.

*Security:* Due to the hash function, cryptographic protocol, and authentication mechanism, the blockchain provides an immutable and secure record for the transaction and a communication medium for sharing the information.

*Anonymity:* The blockchain provides anonymity using the hashing algorithm, which helps to hash the internet protocol (IP) address of the connected node in the blockchain. This prevents users from disclosing their IP addresses and provides privacy in the network.

*Auditability: It* is a crucial feature of the blockchain. In the blockchain, all transactions are recorded with a timestamp, and linking using the Markle tree provides an easy way to audit the whole network. Also, the transparency feature makes the auditing process more reliable.

### 6.2.6 Application of blockchain

*Smart Healthcare:* In healthcare, patient privacy and the security of stored data are very important and crucial. Adopting blockchain technology in healthcare provides privacy and security in the system. The blockchain provides decentralised storage for patient records, which prevents having a single point of failure. Hospitals, healthcare providers, laboratories, pharmacies, and insurance companies are involved [19]. The blockchain provides a secure and direct medium for data sharing among those institutions. Smart healthcare includes advanced technology such as IoT and AI for monitoring and analysis of patient data. This type of environment is developed using multiple sensors and devices that take readings from the patient and send them to the cloud server for processing. These data are crucial for patient treatment and must be authentic and secure while transmitting from device to server. For this, blockchain technology provides a secure and encrypted medium for data transmission and storage [20].

*Supply chain:* Different organisations are involved in product development in the supply chain, from raw material collection to complete products. The process of producing a product from the raw material should be transparent to the consumer, and data integrity must be provided to the consumer. Thus, blockchain plays an important role here as an immutable transaction ledger. This supply chain system is currently advanced with IoT technology that fully automates this process. So, adopting the blockchain in such a system provides decentralisation for processing and storing data in the ledger, which is more convenient than the traditional system. Here, the benefit of the blockchain is that once the product information is written on the ledger, it cannot be changed, and the consumer gets all the information when buying the product [21].

*Smart cities:* Smart cities are developed using the complex architecture of the IoT network, which automatically allows data to be shared within the connected devices. The traditional system of the smart cities architecture includes a centralised server to process the data collected by the IoT device and send it back to improve the performance. The main concern in such a system is the failure of the central server; if the centralised server fails, the smart infrastructure will collapse [22]. Additionally, it is also required to provide security and data integrity at the time of information sharing during device-to-device or device-to-server transactions. The blockchain provides a solution to both issues; the distributed ledger prevents it from a single point of failure, and the cryptographic protocols provide security and data integrity during data communication [23].

## 6.3 Limitations of EI and benefits of blockchain

### 6.3.1 *Limitations of EI*

Although the advancement of EI and 5G on IoT improves the performance of more complex IoT infrastructures such as industry, smart cities, and healthcare, it also faces some specific challenges that affect the adoption of EI in IoT. Figure 6.12 presents the limitations of EI in IoT.

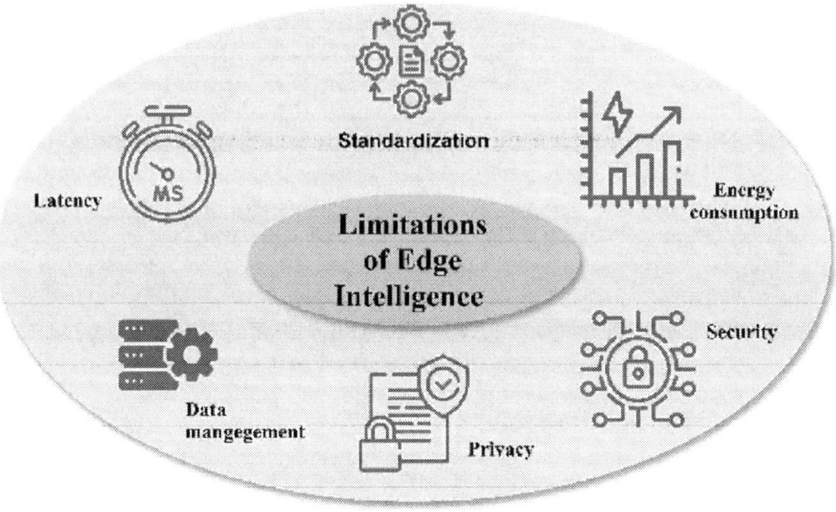


*Figure 6.12* EI limitations in IoT.

*Standardisation:* In the EI, the computation is performed on the edge devices, and these devises can have different computing and energy capacities. The lack of compatibility and coordination among devices can lead to inconsistencies in data processing and communication. To solve such crucial problems a common standard is required to be implemented in the system [24].

*Energy consumption:* Many of the edge devices are running on battery power and they are not compatible to run high computation edge intelligence models. Thus, these EI models can be divided into sub-tasks based on the energy consumption for collaborative execution of the model. However, the creation of this subtask can also lead to the inconsistency between the divided task and the data. Also, it causes insufficient confidence in the result in the case of independent learning of edge devices [25].

*Security:* The openness of edge computing creates vulnerabilities for hardware security. This can be a hardware module, platform, or the software of the edge device. To solve this, a lightweight and distributed security model is required to be deployed for access control and authentication for users and between devices [26].

*Privacy:* The EI is computing data on edge devices that process some sensitive data, such as patient data, and business information related to privacy policies such as European Union's General Data Protection Regulation (GDPR). Such datasets cannot be shared between devices or may cause privacy leakage. Also, the model trained with the real dataset can possess vulnerability to inference attacks against privacy when it is shared with other edge devices [26].

*Data management:* Because of huge data generation and flow over the network, it is difficult to manage it. As many of the users want to protect their data and do not share, this leads to the unavailability of these data for further training and analysis by the model. Additionally, the local data may develop bias in the system for edge computing and it can affect the accuracy of the overall model [24, 26].

Latency: As in the EI, the edge devices work collaboratively and in a distributed manner, which needs to be synchronised. Because of the deployment of the edge devices in different geolocations, there could be delay in model upload and download.

### 6.3.2 Benefits of blockchain

*Standardisation:* The use of blockchain in EI provides standardisation by employing decentralised resource management and database platforms. The blockchain also provides the interoperability in the EI to manage devices with different networks in a decentralised manner [27].

*Energy consumption:* The incentive mechanism of the blockchain can solve these issues, in that the edge devices can share their computing resources against some incentive or rewards. This resource sharing can be done using collaborative and co-operative methods for the computing. This can solve the problem of dividing tasks into sub tasks and removes inconsistency caused by delays in independent learning of edge devices [28].

*Security:* In the blockchain, every node's address is hiding using the hash algorithm and communication is protected by asymmetric encryption. This will secure the communication between devices. Also, the communication between devices is authenticated by the digital signature algorithms. So, if any malicious request for communication is initiated in the network, it will automatically be rejected by the devices [29].

*Privacy:* Because of the hash function of the blockchain, the actual identity of the devices and the users are hidden in the network. Also, the data are encrypted at the time of transmitting from one device to another, which protects the privacy of the data and provides security from unauthenticated access [29].

*Data management:* As the blockchain is a decentralised and distributed system, it reduces the cost of centralised data management and provides data integrity and authenticity using digital signature and timestamp. Also, the cryptographic protocol provides security and privacy to the data at the time of communication between devices or even when it is stored on the device [30].

Latency: The blockchain incentive system can help improve the model's collaborative learning by resource sharing against reward or incentive. In such an environment of collaborative learning, blockchain the authentication process provides trust between the devices. Additionally, the use of smart contract in this type of learning process provides security against poisoning attacks by updating the learning algorithm accepting the participating nodes [30].

## 6.4 Enabling blockchain for edge intelligence for IoT

The adoption of the blockchain can solve the EI problems discussed in the previous section. EI is enhanced using the benefits and features of the blockchain.

Figure 6.13 depicts the layering of the blockchain, which supports improvement in the edge intelligence layers. The blockchain layers relate to the specific issues of the EI layers. Based on the layers and functionalities, we address three main issues in EI: EI model training and optimisation, data management, and power optimisation in computation.

### *6.4.1 Blockchain-enabled EI model training and optimisation*

Deploying the highly intensive EI model on the edge may cause resource constraints and high computation power requirements. Other than this, security-related challenges and inefficiency may also occur. To address these issues, Figure 6.14 presents a blockchain-based network of different IoT devices. The blockchain provides solutions that can be divided as follows:

*Availability:* The blockchain incentive mechanism in the EI motivates the participating edge nodes to share their data with the other edge nodes in the EI model to enhance learning. This can improve the performance and the accuracy of the EI model. The sharing of data and the update in learning of the EI model can be verified and validated by another node in the network. A smart contract-based learning model can be introduced. These smart contracts will help to endorse the EI models with a high reputation, training accuracy, and high performance and share their training dataset to improve the training of the other EI models [31].

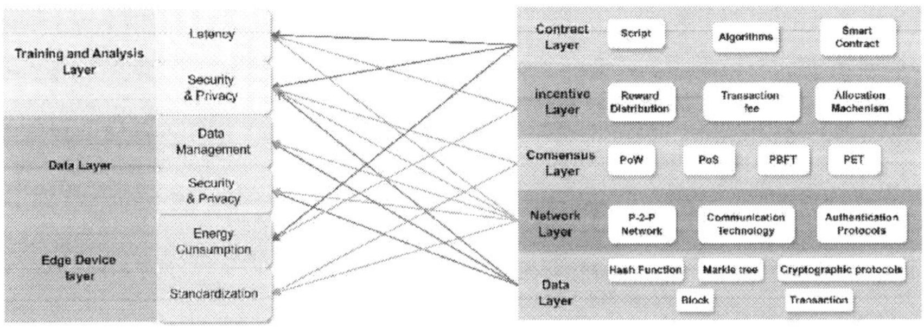


*Figure 6.13* Enabling Blockchain for EI.

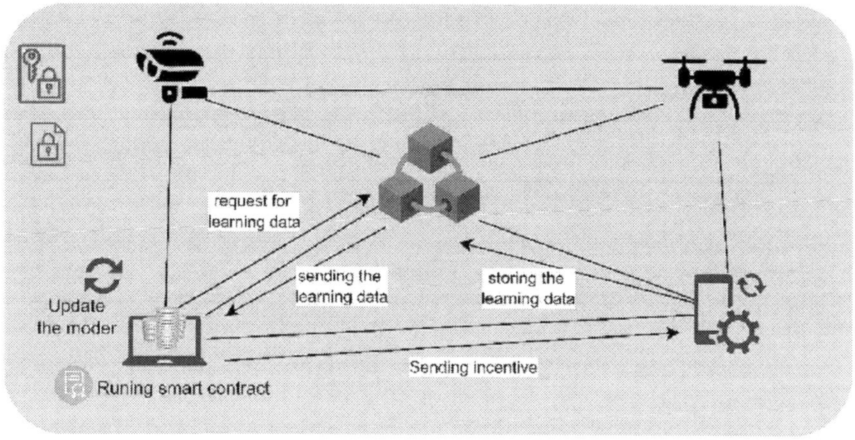


*Figure 6.14* Blockchain-enabled EI model training and optimisation.

*High Efficiency:* In the blockchain-enabled EI, every node can have a copy of the distributed ledger, which supports the training of the model through the previously trained dataset from the other models. The blockchain stores the learning model with the aggregator in the system and later can share it with the other nodes, which helps to achieve high efficiency. In this case, the nodes which are sharing the model will be verified by the other nodes and need for the validation of the model to prevent from the malicious code [32]. The adoption of additional technology, such as federated learning, can improve the edge node learning process and help to achieve high efficiency in the EI model [33].

*Security:* The adoption of distributed and collaborative learning through the shared EI model and dataset possesses many vulnerabilities. To prevent threats, the consensus mechanism of the blockchain selects only those EI models and datasets that are verified as most reliable and highly accurate, which can also be validated by the other nodes. To ensure the security of the EI model, the consensus verifies and validates the model before writing the transaction in the block. The authenticity of the node that shares the EI model in the network can be verified using the public address of the edge node. Also, while sharing the dataset, an encryption algorithm is used to provide protection while transmitting from one edge node to another edge node. The hashing algorithm helps to hide the data and the identities of the node in the ledger at the time of verification and validation of the transaction by other nodes [33].

### 6.4.2 *Blockchain-enabled data management in EI*

*P2P Data sharing:* A large amount of data is processed in the EI model, which contains the sensitive information of the user or company that is crucial to manage. As the blockchain is a decentralised and distributed system, it removes the cost of centralised data management, and data can be shared in a P2P manner from one node to another. Although sharing of data is crucial and it improves the EI models that run on different edge devices, there are some edge nodes that do not want to share the data for collaborative learning. These nodes can be motivated to share the data by incentivising their work [34], as shown in Figure 6.15. Additionally, in the communication of data, sharing also required to maintain the latency. In this process of data sharing, the cryptographic property of the blockchain provides security, authenticity, and integrity to the data. These security and privacy measures include digital signature and timestamp, which help to protect the privacy of users and company information from the malicious attackers or nodes. As the incentive mechanism and the P2P communication trusted environment promote data sharing in collaborative learning, the consensus algorithm helps to reduce the selection process of the data sources by selecting the most appropriate data sources and promoting them on the network [33].

*Data in distributed Ledger:* The data are stored as transactions in the blockchain verified and validated by the consensus. While data are generated and written as transactions, they are protected by the cryptographic protocol of the blockchain from malicious attacks. The privacy protection is done using the hashing algorithm in these data. At the time of accessing the data, the authentication of the request can be verified using the digital signature protocol of the blockchain [35]. However, the EI system can develop some data inconsistency during data generation and analysis in collaboration. This can be addressed by the distributed ledger with the help of the consensus mechanism that verifies and validates by all the

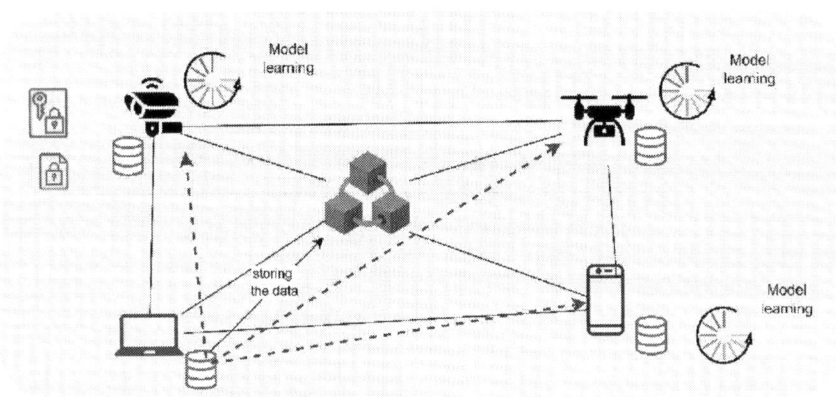


*Figure 6.15* Blockchain-enabled data management in collaborative learning.

connected nodes that the data are successfully written. This consensus can also help to audit the data management process [36].

### 6.4.3 Blockchain-enabled computing power optimisation in EI model

The advancement of edge computing with AI and the processing of large amounts of data require high computation power on edge devices. The power consumption and its management are crucial tasks in edge devices for EI development. To overcome this, blockchain decentralised and distributed systems support the resource-sharing model. In this, the edge devices can share their resources against some benefits, or the resource can be allocated to the device based on the requirement, which is visualised in Figure 6.16. Some of the resource-sharing models are as follows:

*Incentive-based* computing *power sharing:* The blockchain system provides the incentive against the computing power sharing by the node devices, while preserving the privacy of the data. There are two types of nodes: a resource provider and a consumer. So, the consumer (edge device) can rent the computing resources from the service provider. This process requires the verification and validation of the service provider to share power. The incentive generated by the consensus algorithm through this process goes to the service provider [35, 37].

*Performance-based resource sharing:* Based on the specific task of the edge devices, power can be allocated to them. The blockchain smart contract executes the automation process of resource allocation based on the pre-defined task of the edge node [38].

*Computation on Edge server:* In some cases, the size of the data is very large and cannot be processed by the edge devices or be divided as sub-tasks due to the sensitivity of the data. In such cases, an algorithm based on the capacity of the edge computing and the sensitivity of the data can be

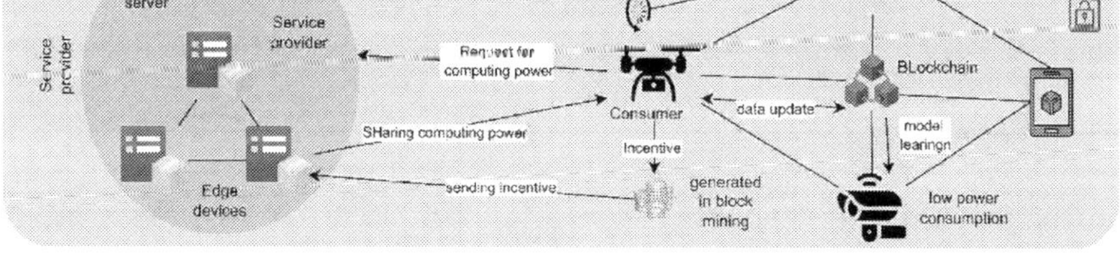


*Figure 6.16* Blockchain-enabled computing power optimisation in EI.

developed as a smart contract to optimise the power consumption of edge devices [39].

## 6.5 Blockchain-enabled EI applications in IoT

### 6.5.1 *Smart healthcare*

The advancement of healthcare technologies has benefited from IoT devices. The wearable technology enables the health monitoring close to the patient and provides accurate information through local analysis. The healthcare infrastructure collects these data in large amounts and shares them within the network to improve the performance of the EI models that are used to analyse those data. The main issue occurs in the EI model at the time of data sharing for collaborative learning of the model. The privacy of the patient data can be compromised by privacy leakage at the time of learning training, which can be reflected in the analysis of the data in that model [40].

The blockchain-enabled EI model provides security and privacy during data sharing and training on the model. The cryptographic properties of the blockchain provide security to the patient data at the time of sharing between the edge nodes, while the hashing mechanism helps to hide the original identity of the patient and the edge nodes. The authentication of the shared data can be verified using digital signature. Additionally, the consensus algorithm verifies and validates shared data [31].

### 6.5.2 *Smart city*

The edge intelligence system enhances data processing and increases the accuracy of the edge devices in the smart city. In the distributed mechanism of EI, the edge nodes such as traffic surveillance camera, security drones, and smart grid generate huge amounts of data. These devices process that data locally and share it with other edge devices to enhance the overall EI model in smart cities [23]. In these devices, information of the individual must be private and sharing the data is challenging in the open distributed network. Additionally, the extensive amount of data creates complexity in analyses at local devices; hence, in such cases the data must be sent to the local edge server to process. This process of data management in the smart city develops concern about the data security and the privacy of the individual in the smart city network [30].

The blockchain-enabled EI model provides security and privacy in data management using the cryptographic protocol of the blockchain. When the data are shared between the edge devices, the data are cryptographically encrypted and cannot be monitored or accessed by any unauthorised nodes. Additionally, the hashing mechanism of the blockchain hides the actual identity (IP address) of the devices during data sharing and hides the individual information inside the data. This provides privacy to the edge devices and to the individuals. Furthermore, the consensus algorithm verifies and validates the data before selecting the appropriate source from the edge nodes. After receiving the data from the edge node, it is authenticated using the edge node's ID in the smart city network [41].

### 6.5.3 *Supply chain*

EI is enhancing the supply chain process from the processing of raw materials to the consumable production and recycling process. In the case of crop production, EI is used in the drone intelligence to monitor crops or spread fertilisers. This drone can process the images of crops with diseases and report the results to connected devices. Then, the transportation of the raw material can be monitored by the smart thermometer inside the vehicle. At the production line, the crop can be classified by the intelligent system based on the quality and packed by robotic hands. These also process the data about the quality of the crops analysed and predict the quality of the whole batch. After the packing is sent to the store, the selling can be monitored by the IoT-based billing system in the store and can predict the demand and popularity of the product. In all these processes, EI models play a crucial role, such that if any of the devices failed, then only those devices need to be checked instead of the whole system [42].

In the blockchain-based EI, supply chain records the data from each operation into the blockchain. This information can be present on the product using a QR code or hyperlink. This process of data sharing gains the trust of the consumer by providing transparency in the production process of the product. This whole process can be automated by using the smart contract of each operation. In the supply chain process, the validation and the verification of the data are performed by the consensus algorithm before writing on the blockchain. The cryptographic protocol of the blockchain provides security to the data, and the digital signature provides authentication in data sharing. Also, the hashing mechanism protects the privacy of the devices and the user's system in the supply chain [43].

## 6.6 Challenges and future directions

*Scalability:* The blockchain can handle the scalability of the IoT network for joining the nodes and running the data analysis and decision-making. However, when EI adopts blockchain for IoT, the main concern that is raised is for edge computing and data generation. Because, in the EI process, data generation is much faster and larger in amount than in the traditional IoT and blockchain system [18]. The blockchain running on a permissionless mechanism has some limitations in throughput, and because of the integration of numerous edge devices that generate huge data, it is difficult to manage the flow of those data and raise the issues of scalability. To address these concerns, further development of an EI network that is able to manage data generation and analysis locally and communicate the optimised data only in the network is suggested.

*Latency:* Because of the advancement in communication technology, such as the adoption of 5G in the IoT network, latency is reduced. Additionally, the adoption of EI improves communication in the IoT network. However, enabling blockchain in the edge intelligence develops the complexity in data flow caused by the consensus mechanism and the smart contracts [33]. Consensus mechanisms and smart contracts in blockchain are dependable protocols on the decision of edge nodes and voting to ensure data privacy and security. This process takes time to reach a common consensus or successful execution of the contract, which takes time to process the data on the network and creates delays in further communication. So, the consensus mechanism is designed in such a way that it reduces the delay in reaching a common consensus and execution of the smart contract.

*Bandwidth:* The adoption of 5G technology enhances the bandwidth of the IoT network and the integration of edge computing reduces the size of data in the communication. The EI models on the edge improve the quality of the competition. However, the adoption of the blockchain in EI makes data processing networks more complex and slower. Blockchain networks are more concerned about the verification and validation of the transactions, which include data from the edge, metadata of the edge device, and execution of the smart contact [33]. This process is handled by the consensus algorithm, and it requires high bandwidth to reach a common consensus in a very short time. In this situation, there is a need for a consensus mechanism to reduce the size of the data. Another solution is using off-chain storage such as an interplanetary file system (IPFS), to store data securely from the outside of the blockchain's main network.

*Intelligence Sharing:* The blockchain incentivises the edge node against data sharing for enhancement of the edge nodes. This helps improve the efficiency of the edge nodes on the network, but the level of understanding of the edge nodes is varied. In such cases the calibration of the shared learning data is very crucial for successful intelligence sharing. Another issue is while sharing the learning data from one edge node to another, it might contain data that may be identified as unnecessary or malicious for other devices. As an intelligent network, it needs to be a developed enough system for understanding the learning data that are shared on the network and optimised for the receiver edge nodes.

*Interoperability:* Although the blockchain provides a secure and privacy-oriented environment in the IoT network, the adoption of EI improves the quality of learning. However, when it comes to cross-border operations, blockchain and EI introduce some difficulties. In the case of blockchain, the protocols that are used to provide security and privacy may differ from one organisation to another, and also from one consensus mechanism to another [44]. To address these issues, a consortium blockchain can be a solution, but when EI is included with blockchain, then it gets more complex. Because of the complexity of the integration of both technologies, the organisations need to develop a system that understands collaborative learning and data sharing without affecting the consortium agreement.

*Social and Environmental Impact:* The advancement and adoption of Intelligent Systems are always in question. Intelligent Systems are self-aware programs that can make their own decision and their use in IoT can manipulate the physical devices, which can harm human beings. Also, people are more concerned about their private data and their privacy in the digital world. Public services are adopting Intelligent Systems that require to verify the user by their social security number. Another concern related to the development of the complex network system is that it demands high consumption of energy. However, the adoption of the blockchain in EI optimise the power in EI, but the consensus used in blockchain uses more power [45]. To this end, a Blockchain-enabled EI must ensure that the advancement of EI does not affect human beings in any negative manner. The data feeding to EI, which contains personal information, must be private and satisfy global regulations like the General Data Protection Regulation (GDPR). And lastly, the development of such blockchain-based EI systems for IoT must be sustainable both economically and environmentally.

## 6.7 Conclusion

In this chapter, we explored EI in terms of its usefulness in the IoT along with the current global situation of IoT adoption and its advancement. We provided a brief introduction to blockchain and later explained it in terms of its fundamental components and their work, the types of blockchains based on the oneness of the network, the properties of the blockchain and its applicability in real time on different sectors. Then we discussed the limitation of the edge intelligence that hindered the wide adoption in the IoT. Based on the limitations, we provided a discussion and explanation of some specific solutions of how blockchain can be enabled in EI to address these issues. The integration of the blockchain in EI improves its performance and other factors like security and privacy of the user, computing optimisation, learning efficiency of the model, and decentralised and distributed collaborative work of edge nodes. However, these benefits of blockchain in EI come with some key challenges that need to be addressed for the successful adoption of these blockchain-enabled EI on in the IoT network. We discussed some of those key challenges along with the future research scope. Although the adoption of the blockchain in EI has benefits and key challenges, this combination of advanced technology helps to develop a future that is secure with widely adaptable systems for the IoT system.

**Acknowledgement:** "This work was supported by the Research Council of Finland with CHIST-ERA, Grant agreement no—359790, Di4SPDS-Distributed Intelligence for Enhancing Security and Privacy of Decentralized and Distributed Systems."

## References

[1] Liu, Y., et al. 2020. "Toward Edge Intelligence: Multiaccess Edge Computing for 5G and Internet of Things." *IEEE Internet of Things Journal* 7 (8): 6722–47. https://doi.org/10.1109/JIOT.2020.3004500.
[2] Sharma, H., and N. Kanwal. 2024. "Video Surveillance in Smart Cities: Current Status, Challenges & Future Directions." *Multimedia Tools and Applications*. https://doi.org/10.1007/s11042-024-19696-6.
[3] Chang, Z., et al. 2021. "A Survey of Recent Advances in Edge-Computing-Powered Artificial Intelligence of Things." *IEEE Internet of Things Journal* 8 (18): 13849–75. https://doi.org/10.1109/JIOT.2021.3088875.
[4] Deng, S., et al. 2020. "Edge Intelligence: The Confluence of Edge Computing and Artificial Intelligence." *IEEE Internet of Things Journal* 7 (8): 7457–69. https://doi.org/10.1109/JIOT.2020.2984887.
[5] Statista. 2023. "Prognosis of Worldwide Spending on the Internet of Things (IoT) from 2018 to 2023 (in Billion U.S. Dollars)." Retrieved September 13, 2024, from https://www.statista.com/statistics/668996/worldwide-expenditures-for-the-internet-of-things/.
[6] GSMA. 2023. "Worldwide Internet of Things (IoT) Connections from 2020 to 2030, by Application (in Billions)." Retrieved September 13, 2024, from https://www.statista.com/statistics/1403256/global-iot-connections/.
[7] Nakamoto, Satoshi. 2008. "Bitcoin: A Peer-to-Peer Electronic Cash System."
[8] Su, X., S. Wang, and R. Yu. 2024. "A Bibliometric Analysis of Blockchain Development in Industrial Digital Transformation Using CiteSpace." *Peer-to-Peer Networking and Applications* 17 (2): 739–55.
[9] Niranjanamurthy, M., B. N. Nithya, and S. Jagannatha. 2019. "Analysis of Blockchain Technology: Pros, Cons, and SWOT." *Cluster Computing* 22 (Suppl. 6): 14743–57. https://doi.org/10.1007/s10586-018-2387-5.
[10] Markets and Markets. n.d. "Blockchain IoT Market Size & Share, Trends Report, [Latest]." Retrieved from https://www.marketsandmarkets.com/Market-Reports/blockchain-iot-market-168941858.html.
[11] Carrara, G. R., et al. 2020. "Consistency, Availability, and Partition Tolerance in Blockchain: A Survey on the Consensus Mechanism over Peer-to-Peer Networking." *Annals of Telecommunications* 75: 163–74. https://doi.org/10.1007/s12243-020-00751-w.
[12] Wu, H. Y., et al. 2022. "Chain or DAG? Underlying Data Structures, Architectures, Topologies, and Consensus in Distributed Ledger Technology: A Review, Taxonomy, and Research Issues." *Journal of Systems Architecture* 131: 102720.
[13] Wang, L., et al. 2019. "Cryptographic Primitives in Blockchains." *Journal of Network and Computer Applications* 127: 43–58.
[14] Kuznetsov, O., et al. 2024. "Adaptive Merkle Trees for Enhanced Blockchain Scalability." *Internet of Things* 27: 101315.
[15] Gupta, M., et al. 2024. "A Comparative Study on Blockchain-Based Distributed Public Key Infrastructure for IoT Applications." *Multimedia Tools and Applications* 83 (12): 35471–96.

[16] Fu, X., H. Wang, and P. Shi. 2021. "A Survey of Blockchain Consensus Algorithms: Mechanism, Design, and Applications." *Science China Information Sciences* 64: 1–15.
[17] Szabo, N. 1996. "Smart Contracts: Building Blocks for Digital Markets." *Extropy: The Journal of Transhumanist Thought* (16): 28.
[18] Wang, X., et al. 2022. "Integrating Edge Intelligence and Blockchain: What, Why, and How." *IEEE Communications Surveys & Tutorials* 24 (4): 2193–2229.
[19] Tanwar, S., K. Parekh, and R. Evans. 2020. "Blockchain-Based Electronic Healthcare Record System for Healthcare 4.0 Applications." *Journal of Information Security and Applications* 50: 102407.
[20] Dewangan, N. K., and P. Chandrakar. 2022. "Patient-Centric Token-Based Healthcare Blockchain Implementation Using Secure Internet of Medical Things." *IEEE Transactions on Computational Social Systems* 10 (6): 3109–19.
[21] Mangala, N., et al. 2024. "Secure Pharmaceutical Supply Chain Using Blockchain in IoT Cloud Systems." *Internet of Things* 26: 101215.
[22] Zanella, A., et al. 2014. "Internet of Things for Smart Cities." *IEEE Internet of Things Journal* 1 (1): 22–32.
[23] Sun, D., et al. 2021. "A Data Stream Cleaning System Using Edge Intelligence for Smart City Industrial Environments." *IEEE Transactions on Industrial Informatics* 18 (2): 1165–74.
[24] Xu, D., et al. 2020. "Edge Intelligence: Architectures, Challenges, and Applications." arXiv preprint arXiv:2003.12172.
[25] Shi, W., et al. 2016. "Edge Computing: Vision and Challenges." *IEEE Internet of Things Journal* 3 (5): 637–46.
[26] Alwarafy, A., et al. 2020. "A Survey on Security and Privacy Issues in Edge-Computing-Assisted Internet of Things." *IEEE Internet of Things Journal* 8 (6): 4004–22.
[27] Salimitari, M., M. Chatterjee, and Y. P. Fallah. 2020. "A Survey on Consensus Methods in Blockchain for Resource-Constrained IoT Networks." *Internet of Things* 11: 100212.
[28] Xiong, Z., et al. 2018. "Cloud/Fog Computing Resource Management and Pricing for Blockchain Networks." *IEEE Internet of Things Journal* 6 (3): 4585–600.
[29] Gupta, R., et al. 2021. "When Blockchain Meets Edge Intelligence: Trusted and Security Solutions for Consumers." *IEEE Network* 35 (5): 272–78.
[30] Manogaran, G., et al. 2021. "Artificial Intelligence and Blockchain-Assisted Offloading Approach for Data Availability Maximization in Edge Nodes." *IEEE Transactions on Vehicular Technology* 70 (3): 2404–12.
[31] Wang, K., et al. 2021. "A Trusted Consensus Scheme for Collaborative Learning in the Edge AI Computing Domain." *IEEE Network* 35 (1): 204–10.
[32] Wang, Y., et al. 2022. "Blockchain Assisted Federated Learning for Enabling Network Edge Intelligence." *IEEE Network* 37 (1): 96–102.
[33] Qiu, C., et al. 2022. "Rendering Secure and Trustworthy Edge Intelligence in 5G-Enabled IIoT Using Proof of Learning Consensus Protocol." *IEEE Transactions on Industrial Informatics* 19 (1): 900–9.
[34] Wang, X., et al. 2020. "Federated Deep Reinforcement Learning for Internet of Things with Decentralized Cooperative Edge Caching." *IEEE Internet of Things Journal* 7 (10): 9441–55.
[35] Lin, X., et al. 2019. "Making Knowledge Tradable in Edge-AI Enabled IoT: A Consortium Blockchain-Based Efficient and Incentive Approach." *IEEE Transactions on Industrial Informatics* 15 (12): 6367–78.

[36] Wang, Q., S. Chen, and M. Wu. 2023. "Incentive-Aware Blockchain-Assisted Intelligent Edge Caching and Computation Offloading for IoT." *Engineering* 31: 127–38.
[37] Lin, X., et al. 2020. "Blockchain-Based Incentive Energy-Knowledge Trading in IoT: Joint Power Transfer and AI Design." *IEEE Internet of Things Journal* 9 (16): 14685–98.
[38] Hu, S., et al. 2021. "Blockchain and Artificial Intelligence for Dynamic Resource Sharing in 6G and Beyond." *IEEE Wireless Communications* 28 (4): 145–51.
[39] Feng, J., et al. 2019. "Cooperative Computation Offloading and Resource Allocation for Blockchain-Enabled Mobile-Edge Computing: A Deep Reinforcement Learning Approach." *IEEE Internet of Things Journal* 7 (7): 6214–38.
[40] Uddin, Md T., G. Zamzmi, and S. Canavan. 2023. "Cooperative Learning for Personalized Context-Aware Pain Assessment from Wearable Data." *IEEE Journal of Biomedical and Health Informatics* 27 (11): 5260–71.
[41] Awotunde, J. B., et al. 2023. "Privacy and Security Enhancement of Smart Cities Using Hybrid Deep Learning-Enabled Blockchain." *Scalable Computing: Practice and Experience* 24 (3): 561–84.
[42] Yin, F., et al. 2023. "Edge Intelligence-Enabled Supply Chain Financial Model Based on Business-to-Business E-Business Platforms." *Concurrency and Computation: Practice and Experience* 35 (13): e6353.
[43] Dedeoglu, V., et al. 2023. "Blockchain Meets Edge-AI for Food Supply Chain Traceability and Provenance." *Comprehensive Analytical Chemistry* 101: 251–75.
[44] Gadekallu, T. R., et al. 2021. "Blockchain for Edge of Things: Applications, Opportunities, and Challenges." *IEEE Internet of Things Journal* 9 (2): 964–88.
[45] Wang, X., et al. 2024. "A Survey on Trustworthy Edge Intelligence: From Security and Reliability to Transparency and Sustainability." *IEEE Communications Surveys & Tutorials* 27 (3): 1729–57.

Chapter 7

# Smart systems powered by IoT with edge computing

*R. Aishwarya and G. Mathivanan*

## 7.1 Introduction

In this chapter, we recognize the crucial roles that edge computing and Internet of Things (IoT) play in today's technology. These technologies are bringing about remarkable modifications across different industries, which offers new probability for data efficiency, data processing, and data connectivity. By traversing the advancement of computing and summarizing the significance of edge computing and IoT, this chapter sets the stage for deeper knowledge about future advancements.

## 7.2 Evolution of computing

In earlier times, mechanical computer tools such as abacus and antikythera were used to support computation. These devices were physically operated and had very limited capabilities. The pivotal development in technology happened in the 1830s with Charles Babbage's analytical engine. Although it was never fully constructed during Charles's lifetime, it laid the groundwork for the programmable computers. The transition from mechanical computing tools to electronic computing happened during the 20th century. In the 1940s, the first electronic computer named vacuum tube (Electronic Numerical Integrator and Computer (ENIAC)) was invented. In the 1950s, transistors were introduced mainly to reduce power consumption and the size of computers by increasing their reliability and speed [1]. In the 1960s, mainframe computers emerged and laid the foundation for business computing. These types of powerful, large systems were able to handle vast amounts of data and multiple tasks simultaneously. IBM-like companies carried out the development of mainframes, which were important for business operations, government applications, and scientific research. The advent of personal computers (PCs) was observed in the 1970s and 1980s, which made computing accessible to individuals and even small businesses. This generation also saw the evolution of graphical user interfaces (GUIs) and the worldwide adoption of software applications. In the 1990s and 2000s, the rise of the

DOI: 10.1201/9781003512066-7

internet brought about the connecting of computers globally and made the exchange of information possible. The emergence of mobile computing with smartphones was also seen in this period. In the late 2000s, cloud computing gained prominence, which allowed users to access computing resources and services over the internet. It offers flexibility, scalability, and cost efficiency. It also provides computing power for businesses and individuals. The essential cloud services include Infrastructure as a Service (IaaS), Platform as a Service (PaaS), and Software as a Service (SaaS).

## 7.3 Significant of edge computing and IoT in contemporary technology

As computing continues to progress, two important modes have arrived: Edge Computing and the Internet of Things (IoT). These kinds of technologies mainly address the constraints of traditional computing models and disclose new innovation and efficiency. Edge computing involves data processing closer to the end user rather than centralized data server. This technique reduces latency, improves real-time data processing, and conserves bandwidth. The important benefits of edge computing include fast response time, enhanced reliability, and privacy and security [2]. It also reduces the operational costs. It is very valuable in applications such as industrial automation, smart cities, and autonomous vehicles which requires immediate data analysis and decision-making.

The network interconnected with physical devices along with sensors, software, and other technologies to transfer and collect data is called the IoT. This kind of service increases the network span for various applications from smart home devices, wearable health monitors to environmental sensors and industries. Also it enables real-time monitoring and enhanced efficiency across various fields and sectors. IoT provides seamless communication, data sharing, and services, which leads to smart and responsive environments during connection between devices and systems.

The combination of edge computing and IoT support each other by labeling the challenges linked with processing and analyzing the vast amount of data generated by the IoT system. Processing data at the edge helps organizations reduce the burden on centralized data centers, lower latency, and improve bandwidth efficiency, thereby enabling timely decision-making. This union is analytical for applications such as remote healthcare, where instant analysis of data can save entities, and in production, where real-time monitoring can protect costly downtime.

## 7.4 Transformation of computing technologies

This section highlights key milestones in the evolution of computing technologies, tracing the transition from mechanical computers to mainframes, and examining the current role of computing in modern business environments.

## 7.5 Early mechanical computer

The examination of computing can be traced back to early mechanical devices planned to aid in calculation. These developing machines set down preliminary work for the enhancement of modern computers.

Abacus: The first and earliest known mechanical computing device, which is used for basic arithmetic operations.

Pascaline: In the 17th century, this mechanical calculator was invented by Blaise Pascal. It performs addition and subtraction.

Difference Engine and Analytical Engine: In the 19th century, this engine was invented by Charles Babbage. It automates the polynomial calculations.

## 7.6 Shift to electronic computers

The conversion from mechanical to electronic computers marked an important cross over in computational speed, efficiency, and capabilities.

Electromechanical Relays: Zuse Z3 computers were used for electromechanical relays for computation in the early 20th century.

Vacuum Tubes: In the 1940s, vacuum tubes were introduced. In the 1940s, vacuum tubes were introduced, as exemplified by machines like ENIAC. They were primarily developed to enhance computational speed and handle more complex calculations. Vacuum tubes were mainly invented to increase the computational speed and complexity.

Transistors: In the late 1950s, the invention of the transistor revolutionized computing by presenting a smaller, energy efficient, and more reliable system than vacuum tubes.

### 7.6.1 *The rise of mainframes*

In the mid-20th century, mainframes emerged as a more powerful, centralized design that can handle large-scale task processing for research and businesses.

IBM 701 and Universal Automatic Computer: Before the availability of mainframes, the mentioned systems were used for business in terms of data processing and scientific calculations.

Batch Processing: It allows multiple jobs to process sequentially without any human interaction. It was introduced by the mainframes.

Time-Sharing: In the modern era of multi-user systems, the advances of mainframe technology have enabled time sharing of systems, allowing multiple users to access the system at the same time.

## 7.7 Computing business in modern era (owning vs. selling)

Computing technologies plays an important role in the concurrent business world in terms of taking strategic decisions, which include owning or selling a company.

Owning a Company: Owning a company requires significant investment in software, hardware, and needed skilled personnel to deal with it all. This kind of choosing to own and maintain one's own company makes benefits such as security, customization and greater control.

Selling a Company: For selling a company leads to enhanced valuation and attractiveness to the probable buyers. A well-organized IT system can reduce costs, provide efficient operations and demonstrate technological competence.

The environment of business computing is evolving continually, with dramatic changes in market demands and emerging technologies.

Cloud Computing: Nowadays many businesses, industries, and organizations are adopting the cloud computing services to reduce the capital expenditure on IT infrastructure. It also offers benefits such as on-demand resource availability and scalability, allowing organizations to respond more efficiently to changing needs.

Big Data and Analytics: The ability to analyze and process large volumes of data in real time is driving data-driven decision-making, enhancing operational efficiency, and enabling the discovery of new business opportunities.

Artificial Intelligence (AI) and Machine Learning (ML): AI and ML are being integrated with business to enhance task automation, improve user experiences, and provide predictive insights.

Cybersecurity: Nowadays cyber threats are more sophisticated; businesses are prioritizing cybersecurity measures to protect sensitive data and maintain trust.

Remote Work: The transfer to remote work due to the Covid-19 pandemic has lead to drastic improvement in virtual collaboration tools, robust networks, and cloud services.

## 7.8 Arrival and impact of cloud computing

### 7.8.1 *Cloud computing—an introduction*

Cloud computing is an important change in the way resources or services are delivered and consumed. Basically, individuals and businesses depend on personal computers or any other local servers to store or process the application. The enhancement of cloud computing, enables users to use and access computing storage, power, and many services [3].

Cloud computing offers several key advantages, as follows:

Scalability: Based on demand resources can be scaled quickly, which provides flexibility and cost effectiveness.

Cost Efficiency: Users can pay only for what they demand and consume of the resources. They can avoid spending large amounts on hardware and software.

Accessibility: Cloud Services can be accessed from anywhere with the internet and facilitate remote collaboration and work.

Maintenance and Updates: Maintenance, security, and software updates are managed by service providers, reducing the operational burden on end users.

## 7.9 Cloud computing services

In general, cloud computing services are divided into three main types: Infrastructure as a Service (IaaS), Platform as a Service (PaaS), and Software as a Service (SaaS). Each type has its own level of abstraction and control that is ideal for different categories of users.

### Infrastructure as a service (IaaS)

This provides virtual computers over the internet. It includes servers, storage, networks, and operating systems, which allows users to hire these resources as needed. It's best for businesses that need to maintain their own apps and data but don't want the hassle that comes with managing hardware.

Examples: Amazon Web Services (AWS) EC2, Microsoft Azure, Google Compute Engine.

Benefits: Superior flexibility, scalability, and control within the compute environment.

### Platform as a service (PaaS)

This offers a higher degree of abstraction by providing an environment made up of hardware, software, and infrastructure necessary for developing, running, or even managing applications. PaaS makes it simpler for developers to concentrate on writing code rather than worrying about what is underpinning everything.

Examples: Google App Engine, Microsoft Azure App Services, Heroku.

Benefits: Simplifies development process; reduces time-to-market; supports collaborative development.

**Software as a service (SaaS)**

Software as a service (SaaS) provides software applications for users over the internet and on subscription basis. Applications are accessible from a web browser without the users having to install and maintain them locally. SaaS has been incorporated extensively into many types of business applications like email, customer relationship management (CRM), financial services, and human-resource processes.

Examples: Microsoft Office 365, Google Workspace, Salesforce.
Advantages: Easy access and reduced software maintenance and costs are predictable with a subscription model.

## 7.10 Cloud computing limitation

### *7.10.1 Evaluating the computing power and memory capacity*

It offers almost limitless computing strength and storage potential, widely termed cloud. It automates the scaling of resources, making it easier to deal with changing workloads without having to buy physical infrastructure. It is very useful for companies who have changing demands or need huge computational power to perform data analysis, machine learning, and other heavy computational tasks.

### *7.10.2 Power consumption*

The capability of the cloud to reduce total power consumption is one of its important benefits. It has the advantage of being energy-efficient due to its design for impressions and exchanges, and using economies of scale on data centers combined with advanced cooling technologies. But the environmental consequences of cloud computing are ultimately determined by what energy sources are used for data centers. Cloud providers adopting renewable energy sources can significantly enhance the sustainability of their operations.

### *7.10.3 Problems of cloud computing*

Despite all its advantages, cloud computing still faces a number of challenges, as follows:

Storing data in the cloud raises concerns about security and privacy, as it may lead to unauthorized access or breaches of confidential information. Therefore, cloud security must be robust, compliant with regulations, and capable of protecting sensitive data effectively.
Downtime and Reliability: Losing control of much infrastructure means that service outages (the more providers you rely on, the more likely this is)

can halt operations. You need to ensure that the providers you select are reliable with strong uptime guarantees.

Vendor Lock-In: Database migrations are complicated, expensive, and can lead to vendor lock-in when you move data out of a cloud provider.

### 7.10.4 *Requirement of server proximity*

That means the demand for servers that are near the point of edge computing, to avoid latency and performance issues. Edge computing, which involves using servers closer to end-users, reduces latency and boosts the performance of time-sensitive applications like online gaming, video streaming, or real-time analytics.

### 7.10.5 *Introduction to cloud-like servers*

Edge servers are essentially in the cloud, but they sit near cloud infrastructure—hence they're often referred to as "cloud-like" and process data at the source. This method helps to reduce both the size of data that is transmitted to centralized Data Centers and also to decrease latency, thereby improving efficiency in processing such data.

Example: Content Delivery Networks are typically used not for data consolidation but to improve performance by deploying localized data centers closer to end-users at the edge, often without requiring new devices.

Advantages: Better response times, less bandwidth usage, and better performance for critical applications.

Cloud computing has changed the way we access and use IT resources, for both organizations and individuals. This provides significant benefits in scal ability, cost efficiency, and access to data, but it also poses hurdles that must be overcome. Edge computing is growing and there are few ways to improve the performance by trying out solutions, such as integration of edge with cloud services, due to increased demand for modern applications.

## 7.11 The Edge of decentralized computing

### 7.11.1 *Edge computing architecture summary*

Edge architecture for data processing moves away from the model of relying solely on a central cloud layer, instead dealing with computational tasks at the edge—nearer to where these requests are being generated in IoT devices or local servers. The key ingredients of this new architectural shift include the following:

Edge Devices: These are the primary data sources creating large amounts of data, like sensors/smartphones/IoT devices.

Edge Nodes: These are the processing units between data source and other parts of network such as gateways, local servers, or edge servers for pre-processing, filtering, reducing, and so forth.

Edge Data Centers: These are regional data centers offering add up power and storage, serving as bridges connecting the edge nodes to no cloud soul-center.

Central Cloud Data Centers: Standardized, centralized data centers that perform large-scale processing for multiple users.

This horizontal hierarchical structure is key to distributing data processing effectively with low latency and reduced bandwidth ensuring that real-time data can be managed, which essentially powers active learning.

## 7.12 Benefits and characteristics of edge computing

Edge computing includes several benefits and characteristics over the cloud [4], as follows.

Less Latency: Edge computing can provide lower latency by operating closer to the data source so the data doesn't require time to travel to a central server and back. This means real-time response is provided.

Saving on Bandwidth: Edge computing can lessen the amount of data sent to central cloud servers through processing and filtering that is performed locally. This lessens the amount of data sent, which can in turn lessen costs for data transmission.

Better Privacy/Security: Edge computing means sensitive data can be processed locally rather than transmitted to a central server, which enhances data privacy and security. In addition, using local processing decreases privacy breaches during transmission.

Better Reliability: Edge cloud computing systems could operate independently of the central cloud if there was a network outage, so the system could continue operating rather than losing some or all of the functionality until the outage is fixed.

Scalability: Edge computing creates scalable solutions through the use of processing power across many edge nodes/devices, so its viable as the number of IoT devices and data sources continues to grow.

### *7.12.1 Processing computation errors*

Approaches that take computation errors in the edge computing architecture into consideration include the following:

Redundancy and Fault Tolerance: Incorporation of redundancy into systems with fault tolerance mechanisms so that, in case there is an error or malfunction, the operation will continue.

Error Detection and Correction: Make use of strong error detection and correction measures to identify computation errors as soon as possible.

Local Backup and Recovery: Maintaining local backups and recovery plans ensures that operations can be quickly restored in case of system failures.

### 7.12.2 *Aspects of response time*

Major factors that contribute to response time in edge computing are as follows:

Distance from Source of Data: The lesser the distance from the source of data, the less the time it takes to transmit data and this increases the response time.

Power of Edge Node Processing: More processing power at edge nodes allows quicker processing of data and response generation.

Network Latency: Reduced network latency through optimal network design and use of routing protocols reduces average response time.

## 7.13 Edge computing platform and case studies with applications

Several significant platforms help include edge computing and offer a variety of tools and services for the same; some are listed as follows:

AWS IoT Greengrass: This is a software that runs on premises to extend the AWS cloud capabilities to local devices allowing data collection, processing, and analysis to be performed at the edge.

Microsoft Azure IoT Edge: Microsoft Azure IoT Edge is a fully managed service that enables the deployment and management of artificial intelligence, Azure services, and custom logic directly on IoT devices.

Google Cloud IoT Edge: Extends the capabilities for AI and machine learning of Google Cloud to edge devices for real-time data processing.

Effective demonstrations and application scenarios of edge computing in the real world include the following examples.

Autonomous Vehicles: With the help of edge computing, real-time data processing can be achieved and it is important for safe and effective operations of autonomous vehicles.

Smart Cities: Edge computing on smart city infrastructure personalizes the city, enhances traffic management, energy distribution, and public safety by analyzing data in real time.

Industrial Automation: This is more efficient and minimizes downtime with the help of predictive maintenance, as well as real-time monitoring and control, within the industrial environment.

## 7.14 Role of IoT in contemporary society

### 7.14.1 *IoT—an introduction*

IoT stands for the network of interrelated physical objects with capabilities to connect and exchange data with other devices over the internet. These devices, typically with integrated sensors, software, and other technologies, can range from simple household appliances to complex machinery and tools. IoT allows for the collection, monitoring, and administration of data in real time, improving efficiency, decision-making, and automation in numerous industries [5].

## 7.15 Control system—the history and evolution

Control systems have been an integral part of modern automation: they are what makes it possible for machines and processes to work as independently and efficiently as possible. The history of the development of control systems has been a keystone in the development of modern IoT technologies.

### 7.15.1 *Aims*

The main aims of any control system are process regulation, maintenance of desired outputs, and optimization. Examples in real time include the following:

Industrial Automation: Control systems govern manufacturing processes to fully control quality and efficiently make use of resources.

Climate Control: A Heating Ventilation and Air Conditioning thermostat maintains selected temperature settings in a timely manner.

Automotive Systems: Advanced vehicles make use of control systems for a number of purposes including the management of the engine, braking systems, and stability control.

### 7.15.2 *Importance to global operations*

Control systems are crucial to the success of operations globally across the world's industries for assurance in the following: Consistency and quality, as through automation, the quality in the output is always maintained; Safety and reliability, since control systems keep track of any occurrence that can lead to failure; Efficiency and cost reduction, as the optimization of processes reduces waste and operational costs.

### 7.15.3 *Quality parameters*

Some quality parameters of control systems include the following:

Accuracy: the degree of exactitude at which the system attains and sustains a specific output.

Stability: Property of the system to come back to its equilibrium if disturbed.
Responsiveness: Time taken by the system to respond as a result of changes in inputs.
Robustness: Quality of achieving a performance under a variety of situations and perturbations.

## 7.16 From computer networks to IoT

The transition from computer networks to IoT is considered a revolution with the integration of control systems with the internet.

### 7.16.1 *Evolution of control system into IoT*

Control systems have evolved over the years from being isolated, stand-alone units to integrated internet-enabled systems. The transition has, therefore, enabled remote monitoring, predictive maintenance, and real-time analytics of data, thereby raising efficiency and responsiveness.

### 7.16.2 *Control system model forms and simulation with IoT*

Some of the IoT control systems in existence include the following:

Centralized Control Systems: One central hub will operate various devices; this is being applied in smart homes.
Decentralized Control Systems: Devices will conduct their operations in a decentralized fashion but will make use of the communication channel for coordination with other devices; in this area, more Industrial IoT applications exist.

IoT and traditional control systems:

> There are many similarities, in that both include gathering and processing data for better decision-making; both strive to automate procedures in order to reduce human intervention and enhance efficiency; and both attempt to work in real time for better adaptation to new, changing conditions and maintaining specific outputs.

## 7.17 Architecture and mechanism of IoT

IoT architecture and mechanisms [6] are intricate, comprising several layers and components set atop one another in order to make things seamlessly work, as shown in Figure 7.1.

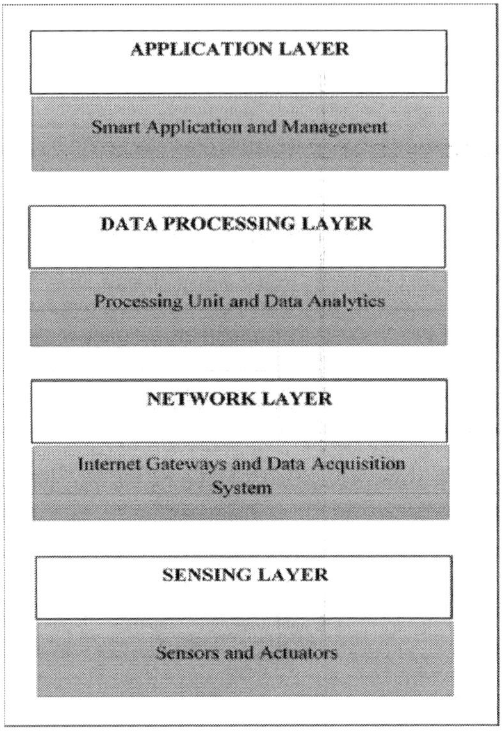


*Figure 7.1* IoT Layers.

### 7.17.1 Integration of sensor

Sensors are fundamental components of IoT, collecting data from the environment. They measure various parameters such as temperature, humidity, light, and motion, providing the raw data necessary for IoT applications as shown in Figure 7.2.

### 7.17.2 Signal process (linear, analog, and digital)

The conversion of data collected by sensors into something that is useful is called signal processing, as shown in Figure 7.3. The main processes involved in this are as follows:

Linearization: Assures that the sensor output is proportional to the measured parameter.

Analog Signals: These are continuous signals that represent physical measurements.

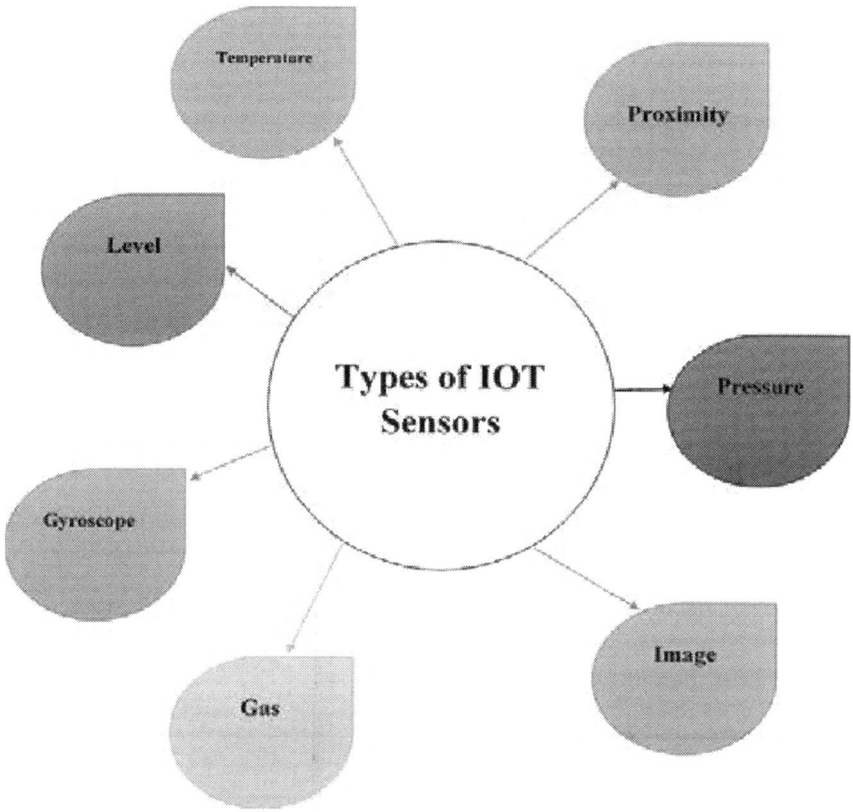


*Figure 7.2* IoT Sensors.

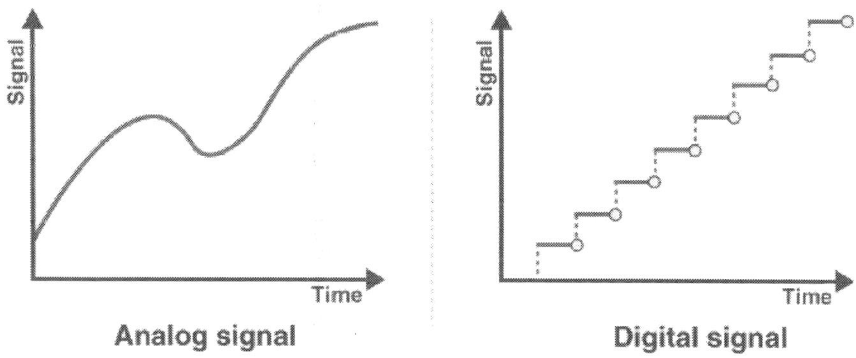


*Figure 7.3* Signal Processing.

### 7.17.3 Logic controller

Controllers will process the signals from the sensor using predefined logic for decision-making. Logic gates in the controller provide basic operations such as AND, OR, and NOT. More complex algorithms can rely on these basic gates and operations for the decision-making process.

### 7.17.4 Continuous action

Continuous action involves making adjustments to aspects of the system in real time to maintain a desired state of the system. The controller will send a command to the actuator based on information received from the sensors, allowing the system to function within predefined values.

### 7.17.5 Actuator and components of IoT

Actuators are devices that execute the control actions, they receive electrical signals and turn them into physical actions. Common actuators we see daily are motors, valves, relays, and so forth. In combination with sensors and controllers, actuators are considered core components of IoT systems, as shown in Figure 7.4.

*Figure 7.4* IoT Actuators and Components.

## 7.18 Embedding IoT with edge computing

### 7.18.1 Why edge computing is needed in IoT

The IoT has grown rapidly and led to a predominant rise in the number of connected devices that generate enormous amounts of data that need processing and analysis. Often powerful traditional cloud computing models struggle with the need for low latency, high bandwidth, and real-time processing required by numerous IoT applications. This is where edge computing comes into play. By processing data closer to where it is generated, edge computing addresses these limitations, enabling more efficient and responsive IoT systems [7].

#### *7.18.1.1 The role of an IoT controller*

In a bid to manage and process data from various sensors and devices within IoT architectures, controllers are very significant. These can be microcontrollers, embedded systems, or even more advanced edge devices equipped with computation power. Their primary functions include the following:

**Data Collection and Preprocessing:** Controllers collect raw data from sensors then perform preliminary filtration, aggregation, and preprocessing operations aimed at reducing cloud traffic volume.

**Local Decision-Making:** Controllers can make real-time decisions and execute actions predicated on predefined rules or AI models diminishing latency as well as improving responsiveness.

**Communication Management:** To achieve effective data transfer and synchronization, they manage communication between IoT devices and the central cloud or other network nodes.

#### *7.18.1.2 System architecture model*

Generally, the architecture of edge computing in IoT is made up of several layers, as shown in Figure 7.5.

**Sensor Layer:** This layer consists of various IoT devices and sensors that generate data.

**Edge Layer:** Edge devices or controllers locally process data. Gateways, microcontrollers, and edge servers are parts of this layer.

**Fog Layer (optional):** These are intermediary nodes that have some additional processing and storage capabilities nearer to edge devices.

**Cloud Layer:** This is a centralized data center where most extensive analysis, storage, and management of data takes place.

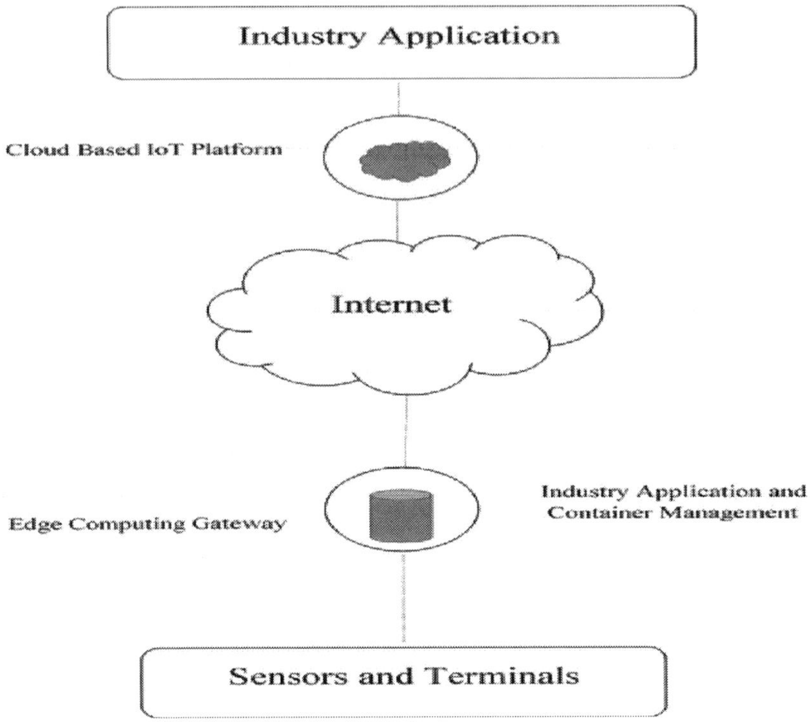


*Figure 7.5* System Architecture Model.

#### *7.18.1.3 Mathematical representation*

Mathematically, edge computing in IoT is represented by models that optimize data communication and processing. The importance consideration is as shown in 1.1, 1.2, and 1.3, as follows:

**Latency minimization:**

$$Latency = \sum_{i=1}^{n} \left( l_{process}(i) + l_{transmit}(i) \right) \tag{1.1}$$

**Bandwidth utilization:**

$$Bandwidth = \sum_{i=1}^{n} \left( B_{edge}(i) + B_{cloud}(i) \right) \tag{1.2}$$

**Energy efficiency:**

$$Energy_{Total} = \sum_{i=1}^{n}(i) + E_{controller}\,(i) + E_{communication}\,(i) \tag{1.3}$$

#### 7.18.1.4 *Advantages and disadvantages of embedding edge computing with IoT*

Edge computing significantly boosts performance and effectiveness of IoT systems through the following:

**Decreasing Latency:** It processes data near the device, thus, making it possible to respond to critical events more quickly, which is very important in cases like self-driving cars and industrial automation.
**Bandwidth Optimization:** It preprocesses data on the edge so that only relevant ones are transmitted to the cloud, which helps in reducing bandwidth usage costs.
**Real-time Analytics:** For time-sensitive applications, edge computing enables people to analyze data immediately as well as make a decision based on such analysis.

### 7.18.2 *Addressing security and privacy concerns*

There are opportunities for edge computing to improve IoT systems' security and privacy, such as the following:

**Data Localization:** Reducing distance between the source and the data consumer decreases the chances of a person or entity intercepting and accessing the data while in transit.
**Decentralized Security:** Installing security measures across multiple edge nodes adds complexity for an attacker trying to compromise the entire system.
**Privacy Preservation:** Sensitive data can be processed at the edge, and only anonymized or aggregated information can be sent to the cloud for processing, reducing the risk of revealing personally identifiable information.

### 7.18.3 *Overcoming deployment challenges*

Even though edge computing offers various benefits, there are a number of deployment challenges, as follows:

**Scalability:** In terms of infrastructure, managing an increasing number of edge devices needs a strong infrastructure and a strong management policy.
**Interoperability:** Establishing empathy between several edge devices, sensors, and the cloud are very complex.

**Maintenance and Upgrades:** Systematic maintenance and edge devices updates are very important for optimal security and performance.

**Resource Constraints:** Edge devices often have limited storage, energy resources, and computing power.

Resolving these challenges will lead to fully leveraging the benefits of edge computing to create more responsive, efficient, and secure IoT systems.

### *7.18.4 Conclusion and future direction*

Edge computing and IoT will lead the upcoming computing generation, which offers uncommon challenges and opportunities. This coaction will enable the enhancement of more responsive, interconnected systems, which will improve business operations and quality of life. By staying updated on these latest developments and exploring them, businesses, infrastructure, and organizations can grasp these technologies to stay innovative and competitive. Devising for the future involves adopting a forward-thinking approach, innovation, and ensuring the representatives can tackle the full potential of edge and IoT in this era. Edge computing and IoT are set to play multiple roles in industries and organizations by permitting real-time data processes, providing solutions to complex problems. The forecast for the technological improvements include advanced processing power, integration of AI and ML, 5G connectivity, powerful security measures, and efficient energy designs. The transformative potential of these technologies includes smart cities, smart healthcare, smart industry automation, smart retail, and smart agriculture. Addressing any technical challenges and the security of edge computing and IoT needs continuous collaboration between academia, government, and industry.

## References

[1] Campbell-Kelly, Martin, William F. Aspray, Jeffrey R. Yost, Honghong Tinn, and Gerardo Con Díaz. *Computer: A History of the Information Machine*. Routledge, 2023.

[2] Kong, Xiangjie, Yuhan Wu, Hui Wang, and Feng Xia. "Edge computing for internet of everything: A survey." *IEEE Internet of Things Journal* 9, no. 23 (2022): 23472–23485.

[3] Yahia, Hazha Saeed, Subhi RM Zeebaree, Mohammed AM Sadeeq, Nareen OM Salim, Shakir Fattah Kak, Adel AL-Zebari, Azar Abid Salih, and Helat Ahmed Hussein. "Comprehensive survey for cloud computing based nature-inspired algorithms optimization scheduling." *Asian Journal of Research in Computer Science* 8, no. 2 (2021): 1–16.

[4] Hossain, Md Delowar, Tangina Sultana, Sharmen Akhter, Md Imtiaz Hossain, Ngo Thien Thu, Luan NT Huynh, Ga-Won Lee, and Eui-Nam Huh. "The role of microservice approach in edge computing: Opportunities, challenges, and research directions." *ICT Express* 9 (2023): 1162–1182.

[5] Das, Sanjana, Jyoti Sekhar Banerjee, Debashis De, Panagiotis Sarigiannidis, Arpita Chakraborty, Mufti Mahmud, and Siddhartha Bhattacharyya. "Role of the Social Internet of Things in Society 5.0: Opportunities and Challenges." In *International Conference on Human-Centric Smart Computing*, pp. 667–686. Singapore: Springer Nature, 2023.
[6] Burhan, Muhammad, Hina Alam, Ahmad Arsalan, Rana Asif Rehman, Muhammad Anwar, Muhammad Faheem, and Muhammad Waqar Ashraf. "A comprehensive survey on the cooperation of fog computing paradigm-based IoT applications: Layered architecture, real-time security issues, and solutions." *IEEE Access* 11 (2023): 73303–73329.
[7] Garcia-Perez, Asier, Raúl Miñón, Ana I. Torre-Bastida, and Ekaitz Zulueta-Guerrero. "Analysing edge computing devices for the deployment of embedded AI." *Sensors* 23, no. 23 (2023): 9495.

Chapter 8

# Edge intelligence–based social robotics

## Planning and its applications

*Kishor Malakar, Rahul Kala, and Vaibhav Malviya*

## 8.1 Introduction

The rapid advancements of robotics and artificial intelligence (AI) are revolutionizing the way we live, work, and interact with technology. As robots and AI systems become increasingly intertwined with our daily routines, they are bringing about unprecedented changes in sectors such as healthcare, education, and entertainment. However, this rapid integration of robots and AI also raises a multitude of complex challenges that require urgent attention. One of the most pressing concerns is the ability of robots to accurately detect and respond to human presence, a crucial aspect of safe and effective human–robot collaboration (HRC). The development of social robots, designed to provide assistance, support, and companionship to humans [1], is a pivotal area of research in this field. These robots are being deployed in diverse settings, including elderly care, healthcare [2,3], and education, to enhance the quality of life for individuals and improve their overall well-being. However, the development of social robots also poses several challenges, such as enabling robots to navigate complex environments, avoid obstacles, and understand human behaviour and intentions. Moreover, social robots must be able to adapt to changing circumstances, learn from experience, and make decisions autonomously, all while ensuring the safety and dignity of humans.

To overcome these challenges, researchers are adopting a multidisciplinary approach that combines mathematical modelling, machine learning, and deep learning techniques. This synergy of technologies has far-reaching implications for various sectors, enabling the development of advanced robots that can predict human motion and intention in real time, taking into account contextual factors such as task requirements, human willingness to cooperate, and environmental constraints. Furthermore, the integration of AI and robotics is enabling the creation of robots that can learn from humans, adapt to new situations, and make decisions based on complex datasets.

This study aims to provide a comprehensive overview of the current state of social robots, including their applications, challenges, and limitations. We will explore the various sectors where social robots are being deployed,

DOI: 10.1201/9781003512066-8

including elderly care, healthcare, and education, and examine the psychological, sociological, and ethical aspects of human–robot interactions. By considering these psychological, sociological, and ethical aspects, researchers and developers can create social robots that are not only effective and efficient but also respectful, responsible, and beneficial to humans. We will also discuss the methodologies being developed to overcome the challenges associated with social robots, including the use of machine learning and deep learning techniques to improve robot perception, cognition, and action. Additionally, we will investigate the potential of social robots to enhance human capabilities and promote inclusivity, particularly for individuals with disabilities or impairments [4].

The potential benefits of social robots are vast, ranging from improved healthcare outcomes to enhanced educational experiences [2,3]. However, the development and deployment of social robots also raise important ethical questions. We will examine the ethical dimensions of robot presence, including issues related to autonomy, agency, and human freedom, and explore the implications of these issues for the development and deployment of social robots. Moreover, we will investigate the role of social robots in promoting social inclusion for People with Disability (PWD) such as the visually or mobility impaired. We will also examine the potential of social robots to enhance human capabilities, such as cognitive and motor skills, and to provide support and assistance in daily life [5].

Modern research in social robotics is filled with the use of deep learning pipelines, where the sensor percept of the robot is given to a deep neural network that makes decisions for the robot, and the decisions are implemented in the environment using the robot actuators. The robots need to act in near-real time, emphasizing short computational times. Doing the calculations on the cloud may lead to high latency, making the system unsuitable for real-time decision-making. At the same time, the robots are often loaded with limited computing on the edge that prohibits using computationally heavy deep learning architectures. As research progresses, it is now becoming important to ensure that the computations can be done within the limited available onboard computer, and only if needed, non-real-time decision-making is done over powerful graphics processing units (GPUs) on the cloud. This is an important aspect to consider in all social robotic applications.

This study will employ a comprehensive review of existing literature to explore the current state of social robots, including their applications, challenges, and limitations. We will also examine the methodologies being developed to overcome the challenges associated with social robots, including the use of machine learning and deep learning techniques. Case studies of social robots being deployed in various sectors will be used to illustrate the potential of robotics to enhance human capabilities and promote inclusivity. Various articles will be studied to identify the key challenges and opportunities in this field. The chapter should be looked into as an organized collection of diverse literature and case studies, highlighting some of the most noteworthy aspects of social robotics. The chapter is not meant to be a stand-alone in-depth coverage of all algorithms and technologies driving social robots, for which the readers are encouraged to refer to the cited papers.

## 8.2 Related work

This section summarizes some interesting works using social robots along with their technologies and challenges. The following sections consider more detailed case studies from diverse use cases. Social robotics, a rapidly growing field, focuses on developing robots that can interact with humans and other robots in a social manner, understanding and responding to human emotions, behaviours, and social cues, enabling effective communication and collaboration. Researchers have made significant contributions, exploring human–robot interaction, social learning, emotion recognition and expression, robot-assisted therapy, and social robot design. For instance, Breazeal and Scassellati's [6] Kismet robot recognized and responded to human facial expressions, while Fong and Simon's [7] robot recognized and responded to human gestures, demonstrating the potential for robots to engage in social interactions. Billard and Dautenhahn's [8] Robota robot learned social behaviours through imitation, and Thrun and Stanley's [9] robot learned through reinforcement learning, highlighting the importance of social learning in robotics. Picard's system [10] recognized human emotions through facial expression analysis, and Breazeal's [11] Leonardo robot recognized and expressed human-like emotions, showcasing the ability of robots to understand and mimic human emotions. Feil-Seifer and Mataric's [12] Bandit robot assisted in autism therapy, and Shibata et al.'s [13] Paro robot provided emotional support to elderly people, demonstrating the potential of robots to improve human well-being. Honda's ASIMO humanoid robot [14] interacted with humans in a social manner, and Sony's Aibo robotic dog [15] provided companionship, illustrating the range of social robots that can interact with humans in various ways. Furthermore, researchers have explored the use of social robots in various domains, including education, healthcare, and customer service, highlighting the potential of social robots to transform various aspects of our lives.

The development of social robots has also raised important questions about the ethics of human–robot interaction, the potential for robots to deceive or manipulate humans, and the need for robots to be designed with social responsibility in mind. As the field of social robotics continues to evolve, it is likely that we will see even more sophisticated and capable social robots that can interact with humans in increasingly natural and intuitive ways. In addition, social robots have the potential to revolutionize the way we live and work, enabling new forms of collaboration and communication between humans and robots. For example, social robots could be used to provide assistance and support to people with disabilities, or to facilitate communication between people who speak different languages. Moreover, social robots could be used to enhance customer service, providing personalized and empathetic interactions with customers. Human social behaviour can also be applied for motion planning and tracking [16–18] and its application is further utilized for tour-guide navigation [19]. In motion planning, it is observed that visibility loss also occurs at the corner point where a robot makes a turn, which can be resolved by behaviour-based finite state machine and behaviour-based tracking [20,21]. The potential applications of social robots are vast and varied, and as the field continues to grow and develop, we can expect to see even more innovative and exciting uses of social robots. Moreover, the development of social robots raises important questions about the future of work and the potential for robots to replace human workers. While some argue that social robots will create new job opportunities and enhance productivity, others worry that they will displace human workers and exacerbate income inequality. As the field of social robotics continues to evolve, it will be important to consider the social and economic implications of these technologies and to ensure that they are developed and deployed in ways that benefit all members of society. Furthermore, the development of social robots also raises important questions about the potential for robots to have their own social relationships and interactions, and the potential for robots to form their own social norms and cultures. As robots become increasingly sophisticated and autonomous, it is likely that we will see the emergence of new forms of robot–robot interaction and social organization. This could have significant implications for our understanding of sociality and social behaviour and could potentially challenge our assumptions about the nature of human–robot interaction. Overall, the field of social robotics is rapidly advancing, and it is likely that we will see significant developments and innovations in the coming years.

## 8.3 Cloud edge computing for socialization of robots based on intelligent data envelopment

The first major challenge in the development of social robots is handling cloud and edge computing to make computing efficient and based on real time for robotic decisions. We discuss the case study [22] as an example of a well-managed cloud edge infrastructure for social robot applications. The study investigates the innovative convergence of cloud edge computing and data envelopment analysis (DEA) to propel the advancement of intelligent social robots. As AI technologies and robotic theories rapidly evolve, the authors in [22] underscore the historical challenges that have impeded the widespread implementation of social robots, particularly their reliance on robust cloud computing infrastructures for optimal functionality. They introduce a groundbreaking framework that merges DEA techniques with cloud and edge computing, aiming to boost the efficiency of robotic systems by minimizing latency, enhancing resource management, and improving user experiences.

Reference [22] offers a comprehensive examination of various DEA models, including traditional efficiency assessments, super-efficiency models, and cross-efficiency models, detailing their application in evaluating the performance of decision-making units within robotic frameworks. This theoretical exploration is paired with practical insights, as the authors present the design of a four-wheeled robot equipped with advanced technologies such as ultrasonic sensors and adaptable mechanical arms. This robot is designed for a range of applications, including underwater environmental monitoring, where it can proficiently gather and relay data on marine conditions.

Additionally, the paper [22] describes the operational structure of the proposed edge cloud computing system, which facilitates the smooth transfer of data between edge clouds, ensuring that users can access essential information with minimal delay. The resource management system is defined by its independence, adaptability, extensibility, and real-time monitoring capabilities, which promote effective coordination among multiple edge clouds. The overarching aim of this research is to refine data analysis processes and bolster the operational capabilities of intelligent robots, thereby enhancing their effectiveness in practical applications. This concludes by highlighting the significance of interdisciplinary approaches in the evolution of intelligent robotic systems, proposing that the introduced framework could lay the groundwork for future developments in the field [22].

## 8.4 Robots as companions and their coexistence

Robots as companions are designed to interact with humans in a safe, social, emotional, and physical way [23]. They have been used in various settings, such as healthcare, education, and rehabilitation, to aid and provide companionship to people with disabilities and disorders. For example, they have been used as healthcare assistants, companions for people with dementia, and exercise coaches.

Research has shown that social behaviour, empathy, and collaboration are important factors in creating a sense of companionship between humans and robots. The way robots are designed and look also plays a significant role in shaping human feelings towards them.

Trust is another important factor that affects how humans interact with robots. When humans trust robots, they are more likely to follow their instructions and are more understanding when robots make mistakes. Trust also helps create an emotional connection between humans and robots, which affects how satisfied people are with the interaction.

Coexistence refers to when two entities share the same space at the same time. Human–robot coexistence is when humans and robots share the same space and interact with each other. This interaction can be meaningful if it feels natural and helps achieve a common goal.

The idea of coexistence leads to the concept of co-performance, where humans and robots work together to achieve a common goal. This can be applied to daily activities, such as doing chores or exercising. As robots become more common in daily life, it's important to find the best way to implement coexistence and co-performance.

Overall, companion robots have the potential to positively impact human lives in various ways, such as improving physical and mental health, enhancing social connections, and providing emotional support. However, more research is needed to fully understand how companion robots can be used in different settings and applications [1].

## 8.5 Domain areas where robots are used as companions

The application of robots as companions has been explored in various domains, with well-being leading the charge. In this field, robots serve as sources of psychological and emotional support [23–26], aid in stress reduction, provide company during isolation, and contribute to depression management. Within the realm of education, robots are used as motivators for reading, act as peer tutors, teach languages, and even provide instruction in sign language. When it comes to socialization, they can reduce loneliness, combat social isolation, and actively encourage social interaction. In healthcare, robots are utilized to assist individuals with dementia, provide domestic health assistance, and contribute to medication adherence.

Beyond these primary areas, robots have found applications in motivation, rehabilitation, entertainment, navigation, guidance, and general assistance. In the context of motivation, they have been employed to encourage exercise, promote healthy eating habits, and influence shopping behaviours. Within rehabilitation, robots play a role in stroke and gait rehabilitation, offer therapeutic support for dementia patients, and assist in walking training. In entertainment, they act as companions for music listening, video watching, and gaming. As navigation and guidance tools, they help individuals navigate both indoors and outdoors, providing directions and companionship along the way. Lastly, in the assistance domain, robots offer socially active support within homes and care facilities.

The widespread exploration of robots as companions across these various domains demonstrates the growing recognition of their potential to enhance human lives in multifaceted ways. Whether offering emotional support, educational assistance, social interaction, or practical help, robots are proving their value as companions in an ever-expanding range of contexts. As technology continues to advance, it is likely that the role of robots as companions will continue to evolve and expand, contributing to a future where humans and robots coexist in increasingly interconnected ways [1].

## 8.6 Hurdles in social robotics

### 8.6.1 *Lack of natural interaction and co-performance*

The current focus is mainly on command-and-response interactions, while more complex and natural interactions like context-aware conversations, facial recognition, and mood recognition are lacking. There's also a need for more studies on scenarios where humans and robots coexist and collaborate naturally [27].

### 8.6.2 *Theoretical and methodological limitations*

The definition of a companion robot remains unclear in some literature. There is also an absence of appropriate frameworks for designing robotic companions and standardized measurement instruments to evaluate HRC [1].

### 8.6.3 *Human-centric focus*

Studies primarily emphasize human perceptions of robots, neglecting the robot's perspective in the interaction. From an action perspective, it needs to be understood that robot kinematics and dynamics will never match that of humans, while from an interaction perspective, the human interaction with a robot will be different to an emotional and interpersonal interaction between two people. Robots are machines that solve problems while being humane for the comfort of humans. Understanding both sides of the interaction is crucial for true companionship [1].

### 8.6.4 *Focus on unusual social circumstances*

A significant portion of HRC research has focused on scenarios like social isolation during the pandemic. While important, this highlights the need for studies exploring HRC in everyday settings [1]. As technologies have shown, it is possible for robots to have autonomous navigation, manipulation, and interaction with humans; research needs to catch up to identifying specific use cases and applications of HRC. Research is boosted by the presence of large-scale datasets, while simply observing human motion and human–human interaction does not solve the problems pertinent to social robotics. This makes it hard to record large-scale datasets in unusual social circumstances, which hampers research.

These limitations present opportunities for future research to explore natural interactions, develop robust theoretical frameworks and measurement tools, incorporate the robot's perspective, and study HRC in broader social contexts [1].

## 8.7 Forward-looking strategies

### 8.7.1 Building from a robot's perspective

It's important to consider how robots perceive interactions and what types of engagement are meaningful to them. Robots could be limited by sensor choices, environment noise models, computing resources available on the edge, latency, kinematic and dynamic restrictions, and so forth. Designing from a robot's perspective can lead to more seamless integration and a better understanding of the dynamics of HRC.

### 8.7.2 Designing for extraordinary or unusual social situations

The Covid-19 pandemic highlighted the potential of companion robots in social isolation scenarios. Future research should investigate how HRC can address other unforeseen social circumstances.

### 8.7.3 Diversity of robot types

There's a need to better understand the relationship between a robot's appearance and type, and its effectiveness as a companion in various scenarios. Comparative studies can reveal why certain types of robots, like anthropomorphic and zoomorphic ones, are more commonly used.

### 8.7.4 Diversity of explored domains and scenarios

While HRC has been widely explored in indoor settings and domains like well-being and education, there's potential in expanding research to outdoor scenarios and other domains like socialization, healthcare, and entertainment. Additionally, research should focus on understanding the domain-specific needs and characteristics of companion robots to improve their effectiveness and efficiency.

These research directions collectively aim to advance the field of HRC by deepening our understanding of the dynamics between humans and robots, expanding the range of scenarios and domains where companion robots can be beneficial, and developing more effective and efficient robotic companions. From the following section, we can see the different methodological and technological agendas on social robotics hurdles [28–32].

## 8.8 Methodological agenda

### *8.8.1 Measurement of HRC*

The current lack of a standardized scale to measure companionship necessitates the development of new instruments that combine various factors, like human-related, robot-related, and interaction-related aspects.

### *8.8.2 Guidelines for designing and evaluating robotic companions*

The absence of comprehensive frameworks for designing and evaluating companion robots calls for domain-specific guidelines that can address unique needs and characteristics across different scenarios [1]. While research has tried to often devise certain social rules that are parametrized by designer-set values, little is known about the human preference for companionship and the hand-crafted social rewards neither offer comprehensive coverage of all social aspects, nor guarantee resemblance to real human preferences.

## 8.9 Technological agenda

### *8.9.1 Efficacy and availability of robotics technology*

Robotic systems need to become more robust and less error prone, especially in outdoor environments. This will facilitate their deployment in diverse settings and expand the scope of their applications.

### *8.9.2 Affordability and accessibility of robots*

To enable mass adoption and study the long-term effects of HRC, companion robots need to become more affordable and accessible to the public [32].

## 8.10 Motion prediction in human–robot interaction tasks

This text presents a comprehensive exploration of the fascinating field of human motion prediction, delving into its significance in both computer vision and robotics, painting a vivid picture of how these disciplines are converging to unlock a deeper understanding of human movement.

Early methods of trajectory prediction considered the ego agent in isolation and predicted the time sequence of ego-agent pose or joint positions based on the past trajectory. Machine learning and neural network models in auto-regressive settings were frequently used. Since the data involve time sequences, Recurrent Neural Networks were of a particular interest. Advances in deep learning enable handling larger dimensional data. For the problem of prediction of navigation, modern approaches often consider the agents neighbouring to the ego-agent whose time-space trajectories are encoded into a latent vector and socially pooled to make the input to a deep learning system. For manipulation and joint human action prediction, the task information along with the participating tools and object information are important parameters to be considered. The use of Graph Convolutional Neural Networks (GCNs) is particularly interesting, where the entire problem is represented as a graph and the future graph states denote the predicted trajectory. Many modern approaches further use Transformer architectures that better model long-term sequences, facilitating a multi-head attention to highlight important aspects of the data, further modelling interaction between all parts of the trajectory to enable making better predictions.

Meanwhile, in the field of robotics, we find a compelling emphasis on the anticipation of human intent, a linchpin in realizing truly seamless and intuitive human–robot interactions. This predictive acumen empowers robots to navigate collaborative scenarios with fluidity, extending beyond basic tasks like handovers to anticipating the complex choreography of multi-person movements. Human intentions, once enigmatic, are now woven into the very fabric of robot planning, paving the way for a new generation of collaborative machines.

Across these interconnected domains, a common thread emerges: a relentless pursuit of ever more sophisticated and granular predictive capabilities. By tapping into the richness of contextual cues, interaction dynamics, and the subtle nuances of human intentions, these methods are ushering in an era where HRC transcends mere technological prowess, becoming a paradigm of natural and intuitive interaction. The ripple effects of this research are vast, promising to reshape virtual reality, animation, and countless other fields where understanding and anticipating human motion is paramount.

As these two disciplines continue their symbiotic dance, intertwining their knowledge and expertise, we stand on the cusp of a profound transformation. Human motion prediction is poised to evolve beyond mere technological advancement, emerging as a powerful tool that can reshape our interactions with the world around us, both digital and physical. The future holds a tapestry of possibilities, woven from the threads of innovation in computer vision and robotics, promising a reality where machines understand our movements as instinctively as we do [33].

## 8.11 D Human motion and intention predictor model

### *8.11.1 Problem definition*

Human motion prediction is one of the most important capabilities of a social robot. We present a case study from the paper [33] that solves the human trajectory prediction problem. Given a sequence of human motion data, represented as,

$$A_{1:N}^{P} = \left[A_{1}^{P}, A_{2}^{P}, \ldots . A_{N}^{P}\right]$$

where each $A_i$ contains the three-dimensional (3D) coordinates of each joint, the goal is to predict future human poses $A_{N+1:N+K}^{P}$ and the underlying intention behind each pose. To add context and make the prediction more accurate, additional information relevant to the specific task is used. This context is represented as $A_{1:N}^{c}$.

For a handover task: The authors focus on the robot's end-effector (REE) and introduce $A_{1:N}^{r}$, which tracks the 3D motion history of the REE. They also include the 3D positions of obstacles in the environment, encoded as $A_{1:N}^{\theta}$.

For a harvesting task: The context involves the positions of grape bunches and the storage box, represented as $A_{1:N}^{g}$, as these are the targets during different sub-tasks. In both tasks, a 'goal intention' is defined, which is a numerical value representing the specific intention the human is likely to express in the predicted future poses. Overall, this approach combines human motion history with relevant contextual information to make more accurate predictions about future human poses and intentions.

### 8.11.2 *Architecture*

The model architecture is shown in Figure 8.1 [33]. The model takes into consideration the REE pose along with the human motion data and obstacle information. Each input passes through a multi-head self-attention mechanism, as widely done in a transformer encoder, while the encodings are pooled together to be used for the prediction. The detected intention is also given as an input to the system. The prediction system uses a GCN architecture. Considering the time series, discrete cosine transform is applied to the data that act as one set of features, while convolution is used to get another set of features, which are together used to make a motion prediction.

### 8.11.3 *Datasets*

#### 8.11.3.1 *Robot harvest dataset*

A new dataset was meticulously crafted in [33] to investigate human motion prediction in the context of human–robot handovers. Ten individuals of diverse ages and genders participated, each performing various scenarios, culminating in 330 unique sequences.

The robot, guided by a predefined path, acted as the receiver, while the human, navigating around strategically placed obstacles, served as the giver.

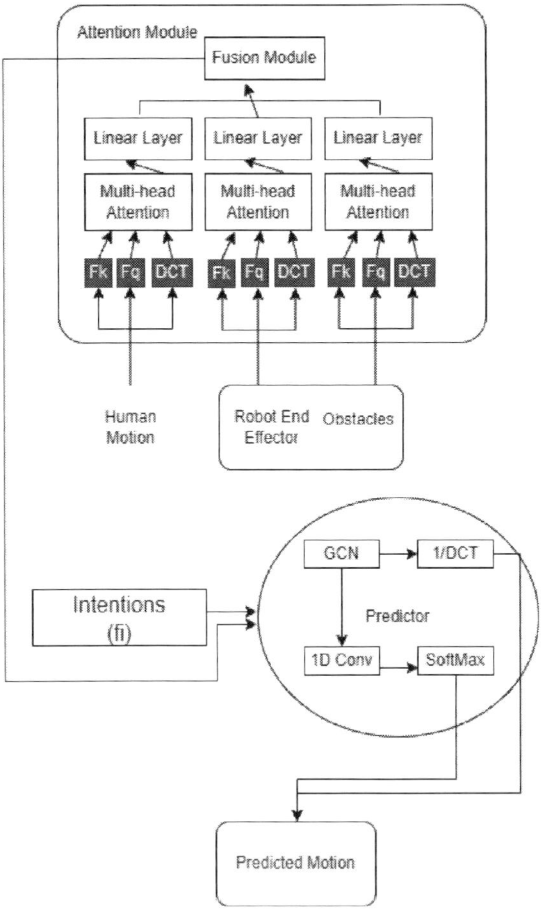


*Figure 8.1* Model overview for the specific case of the Human–Robot Interaction handover [33].

The robot's onboard camera captured human skeletal data, enabling the analysis of their upper body movements relative to the robot. Various obstacle configurations and paths were devised to test how models handle lateral motion and environmental challenges. Crucially, participants were instructed to perform tasks in three distinct ways: naturally, with random gestures, and with deliberate non-cooperation. This deliberate variation in human behaviors aims to illuminate the subtle influence of intentions on motion prediction, allowing models to learn from a broader spectrum of real-world interactions. Each frame was diligently labelled with an intention class to facilitate further investigation into the interplay between human intent and motion during handover tasks [33–36].

#### 8.11.3.2 *Harvest dataset*

A second dataset was compiled in [33] to explore the complexities of HRC in the context of a grape harvesting task. This involved ten participants, spanning various ages and predominantly male, engaging in the task of harvesting grapes alongside a modified mobile manipulator Take-it-and-go (TIAGo) robot designed to navigate the vineyard terrain and carry a collection box. The primary objective was to facilitate the robot's ability to intelligently follow the human carrying the box and efficiently handle box replacements, thereby enhancing the overall collaborative workflow. Furthermore, the dataset seeks to address the critical challenge of human motion prediction in a dynamic, multi-person environment, crucial for ensuring safe and efficient robot operation. Understanding human actions and intentions allows the robot to proactively offer assistance and adapt its behaviour, contributing to a more seamless collaborative experience. Currently, the dataset captures scenarios where the robot remains stationary due to limitations in its navigation capabilities. However, future iterations aim to incorporate robot motion into the data collection process. This inclusion promises to reveal a more comprehensive understanding of the complex interplay between human and robot movements, unlocking deeper insights into the dynamics of collaboration in this context [33].

### 8.11.4 Experiments

The experiments done in [33] are also discussed to understand the overall implications on social robotics.

#### 8.11.4.1 *3D human motion prediction*

In [33], the accuracy of the model was evaluated using L2 distance (i.e., Euclidean distance) between predicted and ground truth sequences, as well as the percentage of frames with errors below certain thresholds. The focus was particularly on the accuracy of the right hand, crucial for the handover task. Incorporating context information, such as the robot's end-effector position and obstacle locations, improves overall accuracy and reduces errors. Intention conditioning enhances intention prediction accuracy, though its impact on L2 metrics was less significant. Importantly, the model can generate diverse trajectories based on different intentions, highlighting its potential for improved HRC.

#### 8.11.4.2 *Handover human–robot validation*

The predictive model was successfully integrated into a robot in [33], enabling real-time predictions of human motion during handover tasks. Initial tests with volunteers showed promising results, although the system currently assumes a cooperative human intention. Future work aims to incorporate an intention classifier for more dynamic interactions. Real-time performance was achieved through parameter adjustments, and evaluations confirmed the model's ability to generate consistent predictions aligned with human trajectories. This successful integration demonstrates the potential of the model for real-world HRC scenarios.

#### *8.11.4.3 User study*

A second dataset was compiled in [33] to explore the complexities of HRC in the context of a grape harvesting task. This meticulously designed study involved ten participants, encompassing a diverse range of ages and predominantly male, actively engaging in the physically demanding task of harvesting grapes alongside a modified TIAGo robot. The robot was ingeniously adapted to navigate the uneven vineyard terrain and equipped with a specialized carrier for the collection box, thus demonstrating the potential for technological innovation to augment and enhance human labour in agricultural settings. The primary objective of this research initiative was to facilitate the robot's ability to intelligently follow the human carrying the laden box and seamlessly manage box replacements, ultimately optimizing the collaborative workflow and minimizing disruptions to the harvesting process. This endeavour not only underscores the significance of robotics in streamlining agricultural practices but also highlights the potential for symbiotic human–robot partnerships to address labour shortages and improve overall productivity. In addition to enhancing efficiency, the dataset aims to tackle a critical challenge inherent in HRC: predicting human motion in a dynamic and often unpredictable multi-person environment. The ability to accurately forecast human movements is paramount for ensuring the safety and effectiveness of robotic operations in real-world scenarios. By comprehending human actions and underlying intentions, the robot can proactively help, adapt its behaviour to accommodate human workers, and ultimately contribute to a more seamless and intuitive collaborative experience. The current dataset, although invaluable in its own right, captures scenarios where the robot remains stationary due to existing limitations in its navigation capabilities. This constraint, however, paves the way for future iterations of the study, where the integration of robot motion into the data collection process holds immense promise. The inclusion of mobile robots will not only deepen our understanding of the intricate interplay between human and robot movements but also shed light on how the robot's own actions influence and shape the dynamics of collaboration. In conclusion, this research initiative, while still in its nascent stages, represents a significant step forward in the pursuit of harmonious HRC in complex, real-world tasks. The insights gleaned from the dataset, coupled with the anticipation of future advancements in robot navigation and motion prediction, paint a compelling picture of a future where robots seamlessly integrate into our workplaces, augmenting human capabilities and transforming the way we approach collaborative endeavours. This research serves as a beacon, illuminating the path towards a new era of human–robot synergy where technology serves as a powerful ally, empowering us to achieve more than we ever could alone [33,37–39].

## 8.12 The robotic chess assistant: promoting inclusivity in competitive chess

We take another case study incorporating social robots from a diverse domain, which is sports. We discuss the approach in paper [4] that offers an interesting application of a social robot.

### *8.12.1 The challenge of disability in competitive sports*

The world of competitive sports often presents significant barriers to individuals with disabilities. Chess, recognized as a sport by the International Olympic Committee, is a notable exception, offering a level playing field for all participants. However, players with severe motor impairments often require human assistance to physically move the pieces, potentially impacting their sense of independence and self-esteem [4].

### *8.12.2 The robotic solution*

The research introduces a robotic assistant designed to address this challenge. The robot, equipped with a robotic arm and an RGB-D camera, can execute moves on a standard chessboard under the player's control. The system employs an intuitive GUI that can be navigated using assistive devices, and a computer vision system to track the opponent's moves. The architecture of the robotic solution is given in Figure 8.2 [4]. The algorithm for this is also given in what follows [4].

### *8.12.3 Key features and advantages*

- **Compliance with Official Chess Rules:** The robot adheres to the regulations set by the International Chess Federation, ensuring its suitability for use in official tournaments.
- **Preservation of Player Autonomy:** The robot is "blind" to the rules of chess, executing even illegal moves if instructed, thus maintaining the player's full control over the game.
- **Intuitive and Accessible Interface:** The GUI is designed for ease of use by individuals with motor impairments and can be controlled using assistive devices.
- **Minimized Time Overhead:** The robot's move execution time is designed to be comparable to that of a human assistant, minimizing disruption to the game flow.

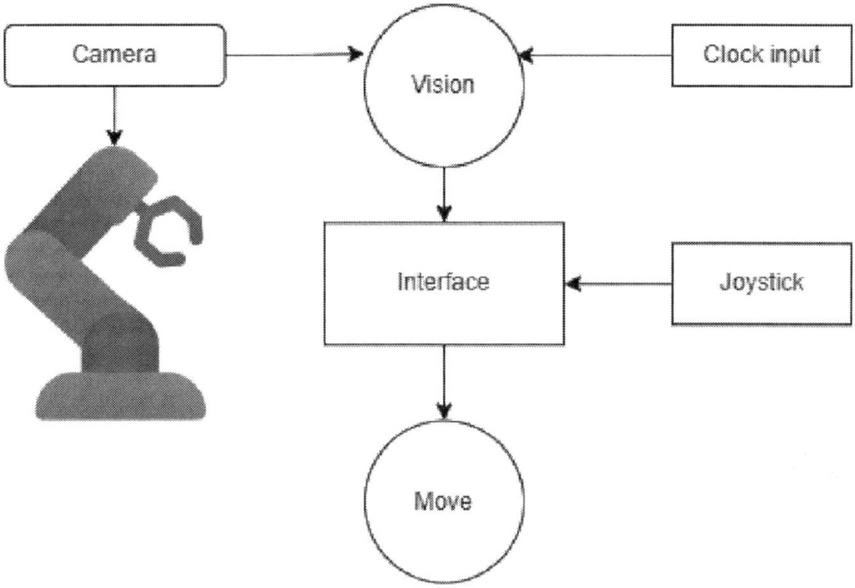


*Figure 8.2* Architecture of Robotic Solution for Competitive Chess [4].

### 8.12.4 *Algorithm*

1: Get map at time *t*-1 from the memory
2: Get the RGB image of the chess board at time *t*-1 from the memory
3: Get the current RGB image of the chess board at time *t*
4: Get the current depth image of the chess board
5: Perform correspondence matching between the RGB images at time *t* and *t*–1
6: Identify which chess grids have changed
7: Based on the old state in the memory identify the move of the opponent player
8: Get the ego-player's move and execute the move.

### 8.12.5 *Experimental validation and future directions*

The system's effectiveness was evaluated in [4] by replicating several famous chess games, with the robot acting as the assistant for one of the players. The results demonstrated a high success rate in move execution and opponent move recognition. Future work will focus on enhancing the system's robustness, speed, and user experience, with the ultimate goal of enabling independent competitive chess playing for individuals with motor disabilities [4].

## 8.13 Bridging the gap: understanding and addressing older adults' needs in assistive robotics

The next case study is taken from the paper [2], where the authors work in the domain of elderly care, understanding the interaction between elderly population and a social robot.

### 8.13.1 *The challenge of user-centric design in assistive robotics*

The development of socially assistive robots (SARs) for older adults necessitates a deep understanding of their needs. However, translating these needs into actionable requirements for research and development poses a significant challenge. Traditional approaches, such as direct questioning or relying on stereotypes, often fail to capture the nuanced and situated nature of these needs [2].

### 8.13.2 *The shortcomings of existing approaches*

The paper [12] critiques existing strategies for understanding user needs, including the following:

- **Loss of Richness and Context:** Methods like affinity diagramming or thematic analysis, while helpful, can lead to a loss of the rich context and qualitative nuances present in the original data.
- **Stereotyping of Users:** The reliance on generalized assumptions about older adults can result in the development of technologies that are stigmatizing and poorly adopted.
- **Lack of Transparency:** The abstraction process from raw data to design implications or personas is often opaque, hindering traceability and validation.

### 8.13.3 *The authentic citations process: a novel solution*

The research proposes a novel "Authentic Citations" process to address these shortcomings. This process involves the following:

- **Identifying the Problem Context:** Clearly defining the research question and the criteria for identifying needs within the qualitative data.
- **Conducting Thematic Analysis:** Systematically reviewing the data to identify and code recurring themes and patterns related to user needs.
- **Developing Themes for Robotics Research:** Organizing the codes into higher-level themes that represent distinct needs and potential areas for robotics research.
- **Developing a Linked Database:** Creating a database that connects the themes, codes, and the original "authentic citations" (direct quotes or descriptions) for transparency and traceability.

#### 8.13.4 Benefits of the authentic citations process

- **Preserves Richness and Context:** Retains the situated and nuanced qualitative data, ensuring a deeper understanding of user needs.
- **Promotes Transparency and Traceability:** Makes the abstraction process explicit, allowing for validation and further exploration of the data.
- **Enhances User-Centricity:** Ensures that the language and perspectives of the users remain central throughout the design and development process.
- **Facilitates Interdisciplinary Collaboration:** Provides a common ground for researchers, designers, and developers to understand and address user needs.

## 8.14 Human-care robot services for the elderly: an experimental evaluation

We discuss another highly related case study taken from [3], where the authors also use a robot for interaction with the elderly population. The authors use verbal and user-specific data identified by the face in addition to the interaction.

### 8.14.1 The growing need for elderly care solutions

The global elderly population is increasing, and the Covid-19 pandemic has exacerbated issues of loneliness and depression among older adults, particularly those living alone. This highlights the urgent need for innovative solutions to provide both cognitive and emotional support to this vulnerable population.

### 8.14.2 The role of human-care robots

Human-care robots offer a promising avenue to address the challenges faced by the elderly. However, existing commercial robots often lack the ability to engage in natural communication, understand individual needs, and foster emotional connections, limiting their effectiveness in providing comprehensive care [3].

### 8.14.3 Developing an integrated human-care robot service

The research focuses on developing an integrated system for human-care robot services that incorporates the following:

- **Personal Profiling:** To recognize individual characteristics, behaviours, and situations.

- **Interpersonal Intelligence Modelling:** To enable natural communication and emotional connection with the elderly.
- **Behaviour Intention Understanding:** To track and understand the daily activities of the elderly.

### 8.14.4 Real-world evaluation

The system was evaluated in the following two real-world environments:

- **Apartment Testbed:** 40 elderly participants interacted with the robot in a simulated living environment.
- **Private Homes:** Two elderly participants living alone used the robot in their own homes for an extended period.
- The evaluation included both quantitative and qualitative assessments.

  **Quantitative:** Measured service success rates and frequency of use to assess stability.

  **Qualitative:** Analysed surveys and interviews to gauge user satisfaction, perception of the robot, and its reliability.

### 8.14.5 Key findings

- **Stability:** The system demonstrated reasonable stability, with higher success rates for on-demand services compared to proactive services.
- **Usefulness:** Users expressed high satisfaction with most services, particularly those offering cognitive and emotional support.
- **Challenges:** Issues such as misrecognition, slow response times, and the need for improved natural communication were identified.
- **Potential:** The study highlights the potential of human-care robots to address the needs of the elderly, particularly those living alone [3].

## 8.15 SARs for at-home cognitive monitoring: a feasibility study

### 8.15.1 The challenge of early MCI detection

We discuss another interesting case study from paper [5], where the authors propose the use of a social robot in the medical domain, which makes an interesting value addition to the chapter. The early detection of Mild Cognitive Impairment (MCI) is crucial for timely intervention and management. However, traditional clinical assessments can be inconvenient and may not accurately reflect real-life cognitive function. The study investigates the feasibility and acceptability of using SARs to administer cognitive tests and monitor cognitive capabilities in the comfort of older adults' homes.

### 8.15.2 *The MoveCare project and cognitive monitoring*

The MoveCare project [5] developed a multi-actor framework that includes a SAR to support independent living through engagement, assistance, and monitoring. The cognitive monitoring module within this framework focuses on the following two scenarios:

- **NeuroPsychological Tests (NPT)**: The robot guides users through digitalized versions of standard neuropsychological tests, providing supervision and support.
- **Spot Questions (SQ)**: The robot asks users single, repeated-in-time questions integrated into their daily activities to enable longitudinal monitoring.

### 8.15.3 *Experimental campaign and results*

As demonstrated by [5], the system was deployed in the homes of ten older adults for at least ten weeks. The robot autonomously interacted with users, administering NPTs and SQs. The results showed:

- **General Acceptance:** Users generally accepted the technology and found it non-invasive.
- **NPT Feasibility:** The robot successfully guided users through the tests, demonstrating the feasibility of at-home administration.
- **SQ Feasibility:** The robot effectively integrated spot questions into daily interactions, enabling longitudinal monitoring.
- **Challenges:** Some technical issues and user preferences were identified, highlighting areas for improvement.

### 8.15.4 *Future directions*

The study [5] provided evidence for the potential of SARs in at-home cognitive monitoring. The system's ability to administer tests and collect longitudinal data offers a promising approach for early MCI detection. Future work will focus on addressing technical limitations, refining the question selection process, and validating the clinical significance of the collected data. The integration of SARs in healthcare settings has the potential to revolutionize cognitive assessment and support for older adults, promoting independent living and timely interventions [5].

## 8.16 SARs and freedom: an ethical and political inquiry

Towards the end of the chapter, we take the last case study highlighting the ethical issues associated with social and assistive robots. Ethical issues are important to consider in any assistive applications, but the ethics associated with robots are seldom studied with much attention. Therefore, the last case study is taken from Reference paper [40].

### 8.16.1 The limitations of the current ethical discourse

The existing ethical discussions on SARs primarily focus on the micro-level of human–robot interaction, emphasizing the potential impact on the well-being and autonomy of individual end-users. This narrow perspective neglects the broader socio-political context in which SARs are developed and deployed, as well as the complex interplay between technology and the structures that shape human freedom. The paper argues that this individual-centric approach fails to address the full normative implications of SARs, particularly concerning issues of justice and the distribution of power.

### 8.16.2 The multifaceted nature of freedom

The study in [40] draws on philosophical insights to unpack the concept of freedom, highlighting its various dimensions:

- **Negative Freedom:** The absence of interference or constraints on individual action.
- **Positive Freedom:** The capacity for self-determination and self-governance.
- **Freedom as Non-Domination:** The absence of arbitrary power or control by others, both at the intersubjective and structural levels.

The study [40] also emphasizes that freedom is not merely an individual attribute but is deeply intertwined with the socio-political structures and power relations within society. It argues that a comprehensive ethical approach to SARs must consider the impact of technology on all these dimensions of freedom.

### 8.16.3 The political dimension of technology

The study [40] highlights the political dimension of technology, emphasizing that technology is not a neutral tool but actively shapes social structures and power relations. It argues that SARs, like other technologies, can have both intentional and unintentional political implications. The development and deployment of SARs can reinforce existing power structures, create new forms of dependency, and shape social norms and values.

### 8.16.4 Reexamining SARs through the lens of freedom

The study [40] applies the expanded concept of freedom to reexamine the ethical landscape of SARs, identifying several key areas of concern, as follows:

- **Power Relations in Care:** SARs can alter the power dynamics between care receivers and caregivers, potentially increasing the risk of paternalism and surveillance.

- **Labour Relations:** The introduction of SARs can impact the working conditions and power relations of care workers, particularly those engaged in "non-nurturant" tasks.
- **Socio-Symbolic Order:** SARs can reinforce social stereotypes and biases through their design and the roles they are assigned.
- **Infrastructure of Human Functioning:** SARs can shape the ways in which individuals convert their capabilities into functioning, potentially limiting their choices and opportunities for self-determination.

### 8.16.5 *Towards a comprehensive ethical approach*

The study [40] concludes by advocating for a more comprehensive ethical approach to SARs that goes beyond the individual-centric perspective. It calls for a critical examination of the political dimension of technology and its impact on the various dimensions of human freedom. This approach would involve the following:

- Considering the broader socio-political context in which SARs are developed and deployed.
- Examining the power relations and structural inequalities that SARs may perpetuate or exacerbate.
- Ensuring that SARs promote human capabilities and opportunities for self-determination, rather than merely providing functional assistance.
- Engaging in open and inclusive dialogue with all stakeholders, including care receivers, caregivers, and technology developers, to ensure that SARs are developed and used in a just and equitable manner.

By adopting this comprehensive ethical approach, we can ensure that the development and deployment of SARs contribute to a more just and equitable society, where technology serves to enhance human freedom and well-being rather than undermine it [40].

## 8.17 Conclusion

The convergence of HRC research, human-care robot services [4], and SARs for cognitive monitoring [40] and ethics has the potential to revolutionize the way we interact with and benefit from robotic technologies. Modern technology, including recent advancements in motion planning and deep learning [41], now enables the use of social robots alongside humans; however, it is important for the robot to understand the human intent and predict the human's actions. A crucial aspect of HRC is the incorporation of contextual information and human intention into motion prediction models, which significantly enhances the accuracy and naturalness of human–robot interactions. This is particularly evident in tasks such as handover and harvesting, where the ability of robots to anticipate and adapt to human actions is critical.

Moreover, the development of robotic assistants, such as those designed for competitive chess [4], can promote inclusivity and autonomy for individuals with disabilities, fostering a more accessible and equitable environment. The Authentic Citations process, which prioritizes rich, contextual, and authentic qualitative data, offers a valuable approach to user-centric design, ensuring that assistive robotics solutions are tailored to the needs and preferences of older adults. In the context of human-care robot services for the elderly, it is essential to address technical challenges, such as improving recognition accuracy and response times, to enhance user experience and acceptance. Furthermore, SARs have shown promise in at-home cognitive monitoring, offering a potential solution for early detection of mild cognitive impairment (MCI). However, future research must focus on refining the question selection process, addressing technical limitations, and validating the clinical significance of the collected data. The ethical implications of SARs cannot be overlooked [40]. A paradigm shift is necessary, moving beyond individual-centric approaches to acknowledge the broader socio-political ramifications of these technologies. By embracing a more comprehensive ethical framework that recognizes the intricate relationship between SARs and human freedom, we can ensure that these technologies are developed and deployed in a manner that truly upholds and enhances human well-being.

In addition, the development of SARs for cognitive monitoring raises important questions about the role of technology in supporting human autonomy and dignity. As SARs become increasingly integrated into healthcare settings, it is crucial to consider the potential impact on human experience, including the risk of dependency, loss of autonomy, and erosion of human relationships [40]. Ultimately, the integration of context, human intention, and user-centric design approaches in HRC research has the potential to create robots that are more capable, intuitive, and readily accepted in various real-world settings. Focusing on technical, ethical, and user-centered aspects paves the way for more meaningful and inclusive robotic applications. Moreover, the development of SARs and HRC systems must be guided by a deep understanding of human values, needs, and preferences. This requires a multidisciplinary approach, incorporating insights from fields such as psychology, sociology, philosophy, and anthropology, to ensure that these technologies are designed to support and enhance human well-being, rather than simply to perform tasks.

In conclusion, the convergence of HRC, human-care robot services, and SARs for cognitive monitoring and ethics has the potential to revolutionize the way we interact with and benefit from robotic technologies. By prioritizing user needs, addressing technical challenges, and fostering ethical discourse, we can create robots that are more capable, intuitive, and readily accepted in various real-world settings, ultimately improving the quality of life for diverse user groups [1–5,33,40].

**Author Contributions:** Authors 1 and 3 carry equal contributions. Author 2 primarily contributed to sections related to trajectory prediction and introduction.

**Acknowledgement:** The work is supported by the National Institute of Technology Mizoram.

## References

[1] Ahmed, Eshtiak, Oğuz 'Oz Buruk, and Juho Hamari. "Human–robot companionship: Current trends and future agenda." *International Journal of Social Robotics* 16, no. 8 (2024): 9–60.

[2] Veling, Louise, and Rudi Villing. "Assistive robotics needs for older care: Using authentic citations to bridge the gap between understanding older persons' needs and defining solutions." *International Journal of Social Robotics* 16, no. 4 (2024): 775–790.

[3] Cho, Miyoung, Dohyung Kim, Minsu Jang, Jaeyeon Lee, Jaehong Kim, Woo-han Yun, Youngwoo Yoon, et al. "Evaluating human-care robot services for the elderly: An experimental study." *International Journal of Social Robotics* 16, no. 7 (2024): 1561–1587.

[4] Pozzi, Luca, Silvia Guerini, Stefano Arrigoni, Alessandra Pedrocchi, and Marta Gandolla. "A robotic assistant for disabled chess players in competitive games." *International Journal of Social Robotics* 16, no. 1 (2024): 173–183.

[5] Luperto, Matteo, Marta Romeo, Francesca Lunardini, Javier Monroy, Daniel Hernández García, Carlo Abbate, Angelo Cangelosi, et al. "Exploring the viability of socially assistive robots for at-home cognitive monitoring: Potential and limitations." *International Journal of Social Robotics* (2024): 1–19.

[6] Breazeal, Cynthia. "Emotion and sociable humanoid robots." *International Journal of Human-Computer Studies* 59, no. 1–2 (2003): 119–155.

[7] Dautenhahn, Kerstin. "Socially intelligent robots: Dimensions of human–robot interaction." *Philosophical Transactions of the Royal Society B: Biological Sciences* 362, no. 1480 (2007): 679–704.

[8] Dautenhahn, Kerstin, and Aude Billard. "Bringing up robots or—the psychology of socially intelligent robots: From theory to implementation." In *Proceedings of the Third Annual Conference on Autonomous Agents*, pp. 366–367. ACM, 1999.

[9] Thrun, Sebastian, Mike Montemerlo, Hendrik Dahlkamp, David Stavens, Andrei Aron, James Diebel, Philip Fong, et al. "Stanley: The robot that won the Darpa grand challenge." *The 2005 DARPA Grand Challenge: The Great Robot Race* (2007): 1–43.

[10] Picard, Rosalind W., Elias Vyzas, and Jennifer Healey. "Toward machine emotional intelligence: Analysis of affective physiological state." *IEEE Transactions on Pattern Analysis and Machine Intelligence* 23, no. 10 (2001): 1175–1191.

[11] Breazeal, Cynthia, Kerstin Dautenhahn, and Takayuki Kanda. "Social robotics." In *Springer Handbook of Robotics*, pp. 1935–1972. 2016.

[12] Fan, Jing. "Models of Adaptation in Intelligent Human-Machine Interaction and Their Applications to Elder Care and Autism Spectrum Disorder Intervention." PhD diss., Vanderbilt University, 2019.

[13] Kang, Hee Sun, Kiyoko Makimoto, Rie Konno, and In Soon Koh. "Review of outcome measures in PARO robot intervention studies for dementia care." *Geriatric Nursing* 41, no. 3 (2020): 207–214.

[14] Kanda, Takayuki, Takahiro Miyashita, Taku Osada, Yuji Haikawa, and Hiroshi Ishiguro. "Analysis of humanoid appearances in human–robot interaction." *IEEE Transactions on Robotics* 24, no. 3 (2008): 725–735.
[15] Sparrow, Robert. "The march of the robot dogs." *Ethics and Information Technology* 4 (2002): 305–318.
[16] Reddy, Arun Kumar, Vaibhav Malviya, and Rahul Kala. "Social cues in the autonomous navigation of indoor mobile robots." *International Journal of Social Robotics* 13, no. 6 (2021): 1335–1358.
[17] Malviya, V., and R. Kala. "Trajectory prediction and tracking using a multibehaviour social particle filter." *Applied Intelligence* 52, no. 7 (2022): 7158–7200.
[18] Malviya, Vaibhav, and Rahul Kala. "Tracking vehicle and faces: Towards socialistic assessment of human behaviour." In *2018 Conference on Information and Communication Technology (CICT)*, pp. 1–6. IEEE, 2018.
[19] Malviya, Vaibhav, and Rahul Kala. "Navigation with a cooperative social robot for a group of visitors using face detection and a 'stop and wait' scheme." *Journal of Scientific & Industrial Research (JSIR)* 82, no. 10 (2023): 1009–1021.
[20] Malviya, Vaibhav, Arun Kumar Reddy, and Rahul Kala. "Autonomous social robot navigation using a behavioral finite state social machine." *Robotica* 38, no. 12 (2020): 2266–2289.
[21] Malviya, Vaibhav, and Rahul Kala. "Socialistic 3D tracking of humans from a mobile robot for a 'human following robot' behaviour." *Robotica* 41, no. 5 (2023): 1407–1435.
[22] Sun, Yu. "Cloud edge computing for socialization robot based on intelligent data envelopment." *Computers & Electrical Engineering* 92 (2021): 107136.
[23] Gross, Horst-Michael, Andrea Scheidig, Steffen Müller, Benjamin Schütz, Christa Fricke, and Sibylle Meyer. "Living with a mobile companion robot in your own apartment-final implementation and results of a 20-weeks field study with 20 seniors." In *2019 International Conference on Robotics and Automation (ICRA)*, pp. 2253–2259. IEEE, 2019.
[24] Jeong, Sooyeon, Laura Aymerich-Franch, Kika Arias, Sharifa Alghowinem, Agata Lapedriza, Rosalind Picard, Hae Won Park, and Cynthia Breazeal. "Deploying a robotic positive psychology coach to improve college students' psychological well-being." *User Modeling and User-Adapted Interaction* 33, no. 2 (2023): 571–615.
[25] Robinson, Hayley, Bruce MacDonald, Ngaire Kerse, and Elizabeth Broadbent. "The psychosocial effects of a companion robot: A randomized controlled trial." *Journal of the American Medical Directors Association* 14, no. 9 (2013): 661–667.
[26] Khosla, Rajiv, and Mei-Tai Chu. "Embodying care in Matilda: An affective communication robot for emotional wellbeing of older people in Australian residential care facilities." *ACM Transactions on Management Information Systems (TMIS)* 4, no. 4 (2013): 1–33.
[27] Jeon, Myounghoon. "Turning HART into HEART: Human emotional AI/Robot teaming." In *Proceedings of the Human Factors and Ergonomics Society Annual Meeting*, vol. 65, no. 1, pp. 1044–1048. Los Angeles, CA: SAGE, 2021.
[28] Roesler, Eileen, Dietrich Manzey, and Linda Onnasch. "Embodiment matters in social HRI research: Effectiveness of anthropomorphism on subjective and objective outcomes." *ACM Transactions on Human-Robot Interaction* 12, no. 1 (2023): 1–9.
[29] Walters, Michael L., Dag S. Syrdal, Kerstin Dautenhahn, René Te Boekhorst, and Kheng Lee Koay. "Avoiding the uncanny valley: Robot appearance, personality and consistency of behavior in an attention-seeking home scenario for a robot companion." *Autonomous Robots* 24 (2008): 159–178.

[30] Fogelson, Donna M., Carolyn Rutledge, and Kathie S. Zimbro. "The impact of robotic companion pets on depression and loneliness for older adults with dementia during the COVID-19 pandemic." *Journal of Holistic Nursing* 40, no. 4 (2022): 397–409.

[31] Ross, Guy Moshe. "Dancing with robots: Acceptability of humanoid companions to reduce loneliness during COVID-19 (and beyond)." *AI & Society* (2023): 1–12.

[32] Wu, Jianming, Donghuo Zeng, Bo Yang, Hattori Gen, Yasuhiro Takishima, Yuta Hagio, Marina Kamimura, Yuta Hoshi, Yutaka Kaneko, and Yusei Nishimoto. "TV-watching companion robot supported by open-domain chatbot "KACTUS"." In *Proceedings of the 20th International Conference on Mobile and Ubiquitous Multimedia*, pp. 230–232. ACM, 2021.

[33] Laplaza, Javier, Francesc Moreno, and Alberto Sanfeliu. "Enhancing robotic collaborative tasks through contextual human motion prediction and intention inference." *International Journal of Social Robotics* (2024): 1–20.

[34] Laplaza, Javier, Albert Pumarola, Francesc Moreno-Noguer, and Alberto Sanfeliu. "Attention deep learning based model for predicting the 3D human body pose using the robot human handover phases." In *2021 30th IEEE International Conference on Robot & Human Interactive Communication (RO-MAN)*, pp. 161–166. IEEE, 2021.

[35] Laplaza, Javier, Nicolás Rodríguez, Jose Enrique Domínguez-Vidal, Fernando Herrero, Sergi Hernández, Alejandro López, Alberto Sanfeliu, and Anaís Garrell. "Ivo robot: A new social robot for human-robot collaboration." In *2022 17th ACM/IEEE International Conference on Human-Robot Interaction (HRI)*, pp. 860–864. IEEE, 2022.

[36] Lugaresi, Camillo, Jiuqiang Tang, Hadon Nash, Chris McClanahan, Esha Uboweja, Michael Hays, Fan Zhang et al. "Mediapipe: A framework for building perception pipelines." arXiv preprint arXiv:1906.08172 (2019).

[37] Garrell, Anaís, Carles Coll, René Alquézar, and Alberto Sanfeliu. "Teaching a drone to accompany a person from demonstrations using non-linear ASFM." In *2019 IEEE/RSJ International Conference on Intelligent Robots and Systems (IROS)*, pp. 1985–1991. IEEE, 2019.

[38] Goldhoorn, Alex, Anaís Garrell, René Alquézar, and Alberto Sanfeliu. "Searching and tracking people in urban environments with static and dynamic obstacles." *Robotics and Autonomous Systems* 98 (2017): 147–157.

[39] Laplaza, Javier, Francesc Moreno, and Alberto Sanfeliu. "Enhancing robotic collaborative tasks through contextual human motion prediction and intention inference." *International Journal of Social Robotics* (2024): 1–20.

[40] Pareto, Júlia, and Mark Coeckelbergh. "Social assistive robotics: An ethical and political inquiry through the lens of freedom." *International Journal of Social Robotics* 16, no. 8 (2024): 1797–1808.

[41] Kala, Rahul. *Autonomous Mobile Robots Planning, Navigation and Simulation*. Academic Press, 2023.

Chapter 9

# Federated learning–based edge intelligence for IoT

*Md Raihan Uddin, Gauri Shankar, and Shareeful Islam*

## 9.1 Introduction

In recent years, the Internet of Things (IoT) has seen rapid development because of a new wave of growth in artificial intelligence (AI) technologies and machine learning (ML) [1]. Notably, IoT devices are increasingly furnished with cutting-edge sensor technologies, robust storage abilities, and powerful processing powers, thereby opening a world of possibilities for connecting various devices to the internet [2]. Nowadays, IoT devices are extensively applied for different data sensing tasks in multiple domains, such as developing smart healthcare, smart grids, smart cities, smart agriculture monitors, and smart industrial control [3]. Effective strategies in these areas should consider the fundamentals of analyzing large-scale data and developing and implementing ML models. These models typically follow the conventional centralized ML processes from cloud servers [4]. However, this approach has significant drawbacks such as latency, high bandwidth requirement, higher energy consumption, and availability. In addition, managing gigantic amounts of distributed IoT applications data, heterogeneous IoT devices, and delivering services at a selected performance and requirement with cloud infrastructure seems complicated [5, 6]. Also, some IoT application scenarios require real-time responses, robust behavior, and higher accuracy, which is almost difficult for cloud architecture to provide [7].

Moreover, a critical issue is the security and privacy implications of uploading and transmitting clients data to a central server for global model building [6]. Edge Computing is a decentralized computing framework that can address the storage, computing, and application of IoT data acquired by edge IoT devices. It focuses on doing these tasks in close contact to the data source (IoT layer) [8, 9]. This new edge computing paradigm, also known as edge intelligence, provides real-time responses, robust behavior, and higher accuracy, thereby instilling confidence in its capabilities to meet the needs of IoT applications [10, 11].

Edge intelligence can be used on smart devices with attached sensors and devices available at gateways near the sensors. For example, wearable smart

DOI: 10.1201/9781003512066-9

devices like smartphones or smartwatches and gateway devices like microcontrollers can serve as edge nodes [5]. Nevertheless, the main challenge with this approach is the significant resource requirement for implementing, training, and updating results parameters on IoT devices with restricted computing, energy, and storage capabilities. Furthermore, very small data samples and local variations in data can lead to an even more flawed model [6]. To solve the problem, FL is a perfect methodology for addressing distributed data and model training. The collaborative training of a standard shared model is facilitated by aggregating and averaging updates computed locally by IoT devices [12]. As a result, FL can achieve an excellent global model, and user data privacy is the utmost priority [12]. Figure 9.1 illustrates some examples of domains with applications that directly benefit from the FL-based edge intelligence in IoT architecture.

This chapter' primary purpose is to gather and introduce recent research focused on trending technology and all the features of federated learning,

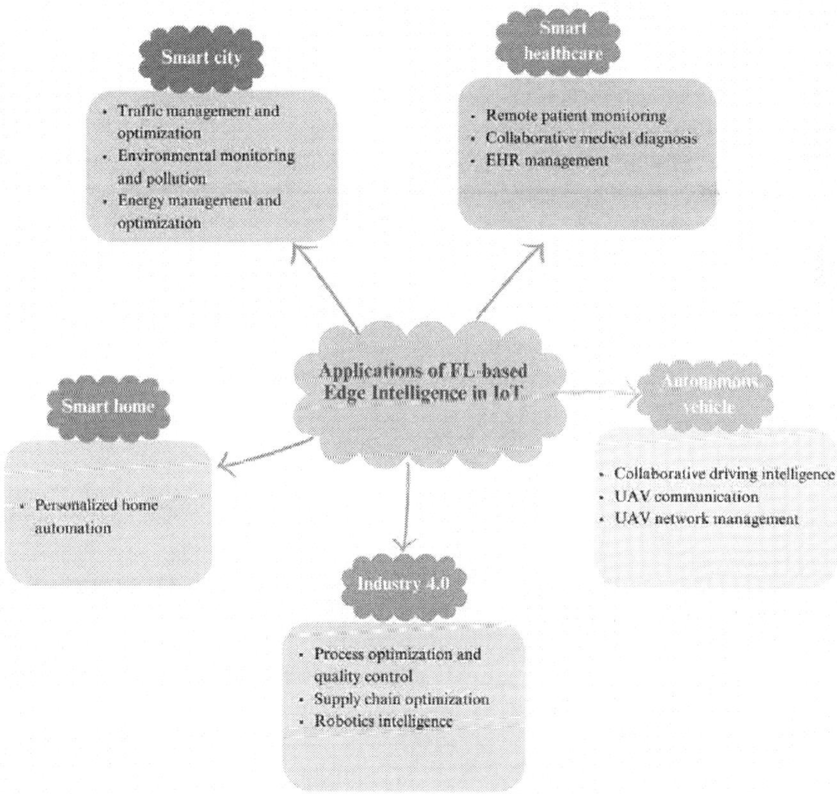


*Figure 9.1* Real-world FL-based edge intelligence in IoT applications with domain.

which is used extensively to improve IoT applications and their services with the assistance of edge intelligence. We discuss how edge devices can collaboratively learn and update models using the FL technique within an edge intelligence IoT architecture. We provide practical examples and scenarios to demonstrate how this framework improves real-time decision-making and resilience while maintaining privacy.

The rest of the chapter is as follows: Section 2 examines the fundamentals of FL, where we present the characteristics of FL, FL architecture, categorization of FL, Communication, and aggregator protocols of FL. In Section 3, we discuss the integration of FL with edge intelligence in IoT, where we present the role of edge intelligence in IoT, FL as a privacy-preserving framework, where we discuss FL-based edge intelligence architecture in detail, and application of FL-based edge intelligence in IoT. Section 4 discusses challenges and future directions, and finally, in Section 5, we present the conclusion with a summary of the chapter.

## 9.2 Fundamentals of federated learning

> *Federated Learning is a machine learning setting where the goal is to train a high-quality central model while keeping the training data decentralized. Instead of bringing the data to the model, the model is brought to the data, trained locally on individual devices, and only the model updates are sent back to the central server. This process is repeated across multiple rounds, enabling the model to learn from a diverse set of data. Importantly, this is all done without compromising user privacy, making federated learning an ethical and user-friendly approach.*
>
> [13, p. 1]

### 9.2.1 *Characteristics of FL*

#### 9.2.1.1 *Decentralized data training*

In conventional ML, data from many sources are gathered in a centralized server for the purpose of training the model. However, FL significantly changes this approach by enabling models to be trained directly on decentralized data. This means that the raw data stay only on local devices like smartphones, IoT devices, or organizational servers, leading to a more efficient model training process. Moreover, these data are inaccessible and remain confidential, ensuring improved ML efficiency [14]. Here, there are two possible ways of data distribution between the clients.

- Homogeneous or independent and identically distributed (IID): This scenario assumes client data distribution is IID. Every client's data are sampled from the identical same feature and statistical distribution. As a result, the data from different clients are statistically similar, which makes it easier to combine local models [15].

- Heterogeneous or non-independent and identically distributed (non-IID): This scenario assumes that the data distribution across clients is non-IID, implying that each data follows a distinct probability distribution [16]. Data heterogeneity in FL can be categorized into the following three types:

  (1) **Feature Space Heterogeneity:** The clients' data have different features yet share a common learning objective.
  (2) **Label Space Heterogeneity:** The input spaces of different clients are comparable, but the label space varies based on each client's specific data characteristics.
  (3) **Combined Feature and Label Space Heterogeneity:** Differences in the feature and label spaces across clients' data lead to significant variability in local models.

#### *9.2.1.2 On-device model training*

One of the primary purposes of introducing FL is the on-device model training approach. This method does the local training of each local ML model using device data, which we call data owner nodes [17]. The locally trained learning models generally have a shared architecture. Every device uses its local data to compute model parameter updates. Following local training, only the computed model changes are transmitted to the central server, while the actual data remain unaffected. Subsequently, the server consolidates these updates to establish the global model as a benchmark for accuracy. The learning process continues to make the model more precise. This approach ensures that confidential information remains stored on the device, thus minimizing the possibility of data breaches [15, 18].

#### *9.2.1.3 Communication efficiency*

Communication efficiency is crucial in FL as the training process involves numerous rounds of interaction between a central server and numerous devices. FL minimizes transmission overhead by transmitting only model updates as there is no need to transmit the whole training dataset [15]. This is achieved through various techniques. Some approaches are model compression, clustering, sparsification, asynchronous updates, and importance-based updates. These techniques are mostly used to minimize the volume of data sharing [19, 20]. Therefore, aggregating and communication protocols play a significant role as they provide effective and efficient operation of FL systems across diverse and resource-constrained settings [21].

#### 9.2.1.4 *Heterogenous devices*

One feature of FL is it can manage the challenges of different kinds of devices with different computing capabilities, storage capacities, and network conditions. In an FL environment, smartphones, tablets, IoT devices like smart thermostats, fitness trackers, industrial sensors, and many other types of devices contribute to training a global model. These devices can vary in their hardware specifications, for instance, processing power, battery life, storage capacity, and connectivity reliability. FL has developed to address these challenges [18]. FL allows each device to participate according to its capabilities. There are algorithms, frameworks, and different aggregator protocols that dynamically adjust the workload allocated to each device. Thus, FL manages all the heterogeneous devices that can contribute to the model training process per their capacity [20, 22].

#### 9.2.1.5 *Cross-organizational FL*

The original FL concept has been extended further beyond its framework. In traditional FL, same feature data use to train the model with different sample size. However, FL can train the model for the different feature data sample. This transformation underlines the high value of data privacy in collaborative ML, confirming that data is always secure [15].

This extension of FL allows cross-organizational collaboration by enabling different enterprises to participate in an FL framework while maintaining data privacy. For instance, a bank holds data of a client who is purchasing using their cards. An e-commerce platform holds data about their products. Through this collaboration, both parties can develop a collaborative model that uses multiple data sources and contains different feature dimensions. This model increases accurate and personalized product recommendations while assuring sensitive information remains protected [16]. These advantages make FL practical in sectors where collaboration across organizations needs to get results from data-driven insights, yet data privacy is of utmost concern.

### 9.2.2 FL architecture

The most common architectures in FL are the client–server model and peer-to-peer (P2P) network architecture. Every architecture has different characteristics, advantages, and challenges when managing data, models, and communication in the FL setting [23].

#### 9.2.2.1 *Client–server architecture*

In this architectural framework, a central server manager node manages the entire learning process. The server node starts the process by sharing a global model with all participating nodes. We call every node a client. Afterwards, each individual client utilizes their own data to train the local model. This ensures the data remain stored on the client's device. After each round of local training, every client sends their model updates as parameters or gradients to the central server. These updates optimize the learning process of the global model and make the necessary adjustments to improve the model's accuracy [15]. This architecture's ability to reduce transmission overhead between clients and the server by just switching model changes rather than raw data is one of its benefits. Google's development of Gboard for Android is an excellent example of this architecture, where the system learns from user interactions without centralizing sensitive data [23]. Nevertheless, the client–server architecture depends completely on the central server, and it creates a single point of failure. Figure 9.2 illustrates the overall working process of a client–server FL architecture.

#### 9.2.2.2 *Peer-to-peer network architecture*

In this architectural framework, there is no central server. The model training and aggregation are distributed across all participating nodes. Each client independently manages its own model and works closely with neighboring clients. This architecture provides secure communication between clients. It does not need a third-party coordinator. After each clients model training using their own data, they share parameters with neighbor clients

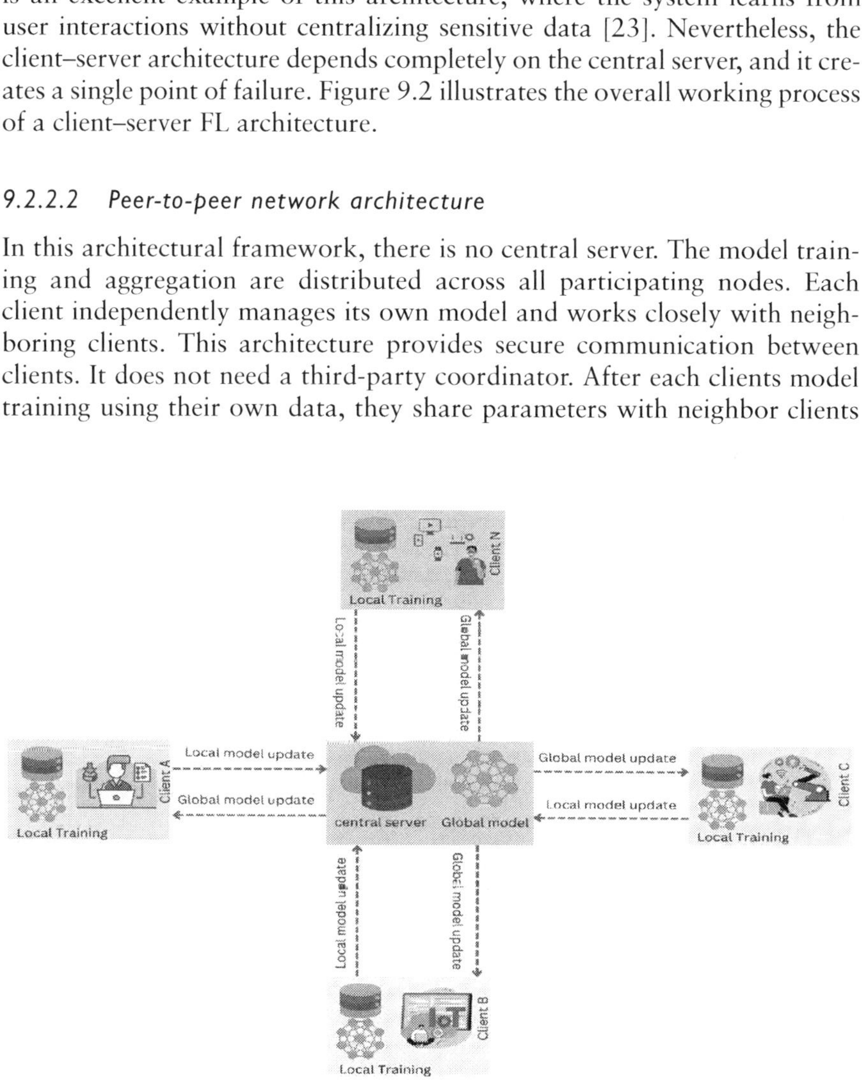


*Figure 9.2* Client–server FL architecture working process.

to make a global model. Thereby, there is minimum risk of data breaches as there is no central server [15, 23]. However, this approach makes high computational demands on each client due to the necessity to encrypt and decrypt messages during communication. This results in the complexity of implementation and higher communication because each client must manage interactions with multiple peers. Overall, the architecture's robustness, scalability, and resistance to single points of failure make it a secure and viable option, particularly in environments where foremost security is a priority [1]. Figure 9.3 illustrates the overall working process of a P2P client–server FL architecture.

When choosing between client–server and P2P architectures in FL, it is suggested to consider the application's specific requirements. Some of the requirements could be scalability, security, computational cost, latency, and bandwidth issues. The client–server architecture is mostly used because of its simplicity and lower communication cost. In contrast, P2P architecture is useful in scenarios requiring high privacy and security. A thorough knowledge of these architectures helps practitioners to design and choose FL systems based on their application scenario.

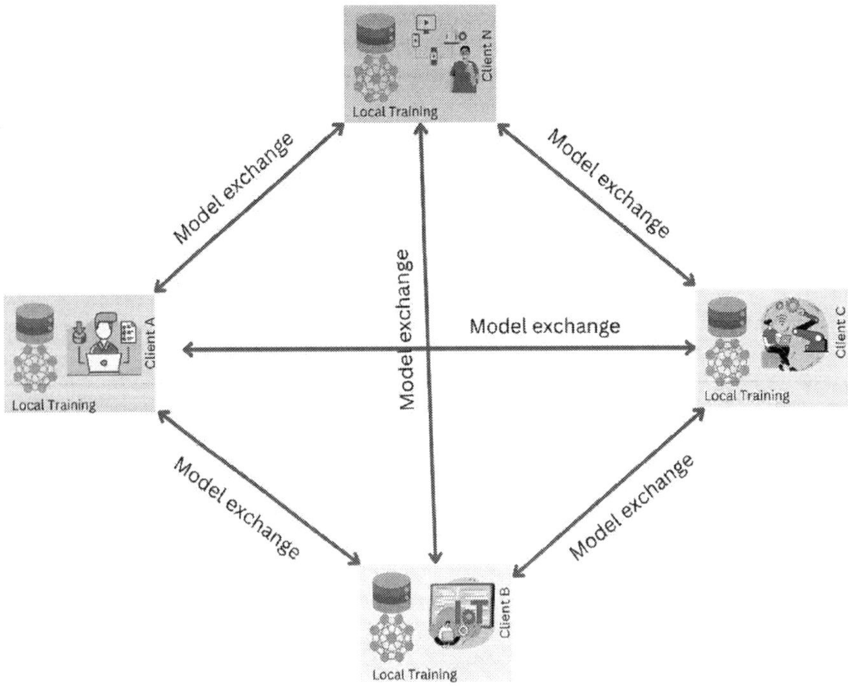


*Figure 9.3* P2P network FL architecture working process.

### 9.2.3 *Categorization of FL*

FL is classified into horizontal FL, vertical FL, and federated transfer learning (FTL). This categorization is based on training data distribution across the feature space and sample space.

#### 9.2.3.1 *Horizontal FL*

Horizontal FL (HFL), also known as homogeneous FL, performs in an environment where the training data are provided by different devices, and the features of the data are the same but the data sample size is different [24]. This method is especially useful when training models on data gathered from similar devices; for example, smartphones or IoT devices, with the same features across clients but varying dataset size [25]. Formally, this is expressed as follows:

$$X_i \neq X_j, Y_i = Y_j, I_i \neq I_j, \forall D_i, D_j, i \neq j$$

where *Xi* and *Xi* contain the feature spaces and *Yi* and *Yj* contain the label spaces for clients *i* and *j*. *Di* and *Dj* represent data of the clients *i* and *j*. The feature and label spaces are identical across clients, but the samples *Ii* and *Ij* differ [15]. Figure 9.4 illustrates the working process of HFL.

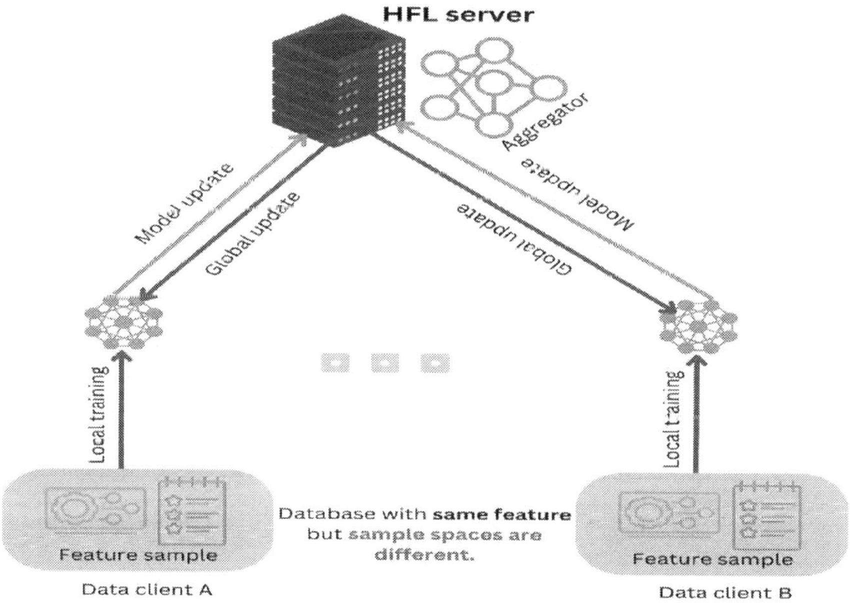


*Figure 9.4* Horizontal FL working process.

#### 9.2.3.2 *Vertical FL*

Vertical FL (VFL), also known as heterogeneous FL, performs in an environment where different devices or organizations hold different features for the same set of data samples. This method is useful when organizations or devices hold information about the same set of users or entities but their features are completely different [24]. Data from different clients are connected by matching the entities (e.g., users or customers) across datasets. This allows the training of a model without sharing raw data. This approach is useful in scenarios where privacy concerns are an utmost priority related to direct data sharing. It allows model training on a different feature set without compromising the confidentiality of the individual datasets [25]. Formally, this is expressed as follows:

$$X_i \neq X_j, Y_i = Y_j, I_i = I_j, \forall D_i, D_j, i \neq j$$

where $Xi$ and $Xj$ contain the feature spaces, and $Yi$ and $Yj$ contain the label spaces for clients $i$ and $j$. Here, $Di$ and $Dj$ indicate dataset of the clients $i$ and $j$. The feature space differs across the clients, but the samples and shared label spaces $Ii$ and $Ij$ are the same [15]. Figure 9.5 illustrates the working process of VFL.

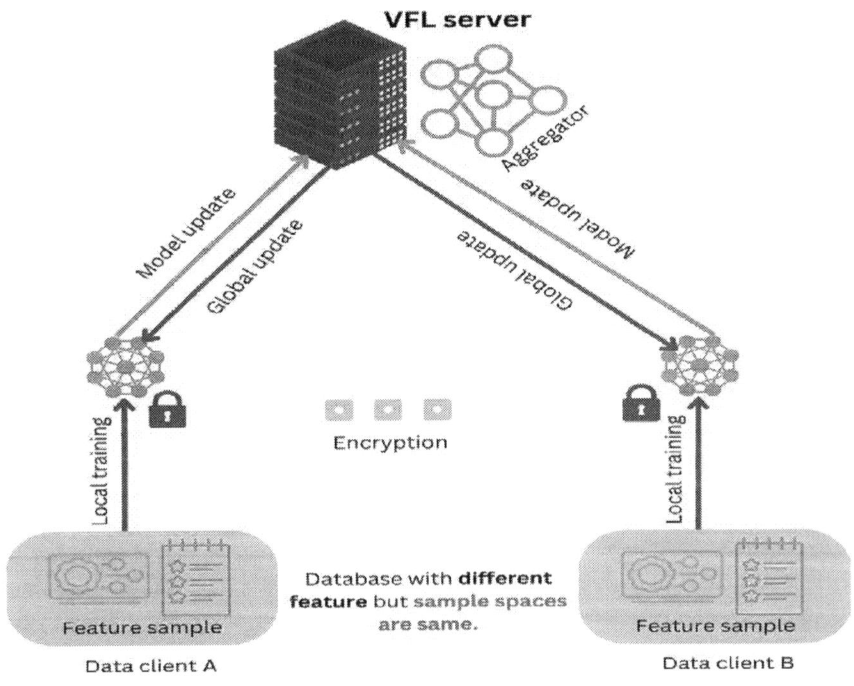


*Figure 9.5* Vertical FL working process.

#### 9.2.3.3 *Federated transfer learning*

Federated transfer learning (FTL), also known as cross-domain FL, performs in environments where clients contain different feature spaces, level space, and different data samples. This method is useful when organizations or devices have extremely heterogeneous data, but it aims to collaborate on building a model [24]. It provides knowledge sharing across clients with non-overlapping feature spaces and sample sets, building effective model training in heterogeneous data environments. This method is useful for the application of intelligent manufacturing, medical treatment, graphics, and image processing, as data heterogeneity is high in these environments [16]. This results in the sharing of knowledge from one client to another, integrating valuable insights from different datasets into a global model [25]. Formally, this is expressed as follows:

$$Xi \neq Xj, Yi \neq Yj, Ii \neq Ij, \forall Di, Dj, \text{i} \neq j$$

where $Xi$ and $Xj$ contain the feature spaces. $Yi$ and $Yj$ contain the label spaces for clients $i$ and $j$. Here, $Di$ and $Dj$ indicate dataset of the clients $i$ and $j$, and the feature space, level space, and data samples are all different [15]. Figure 9.6 illustrates the working process of FTL.

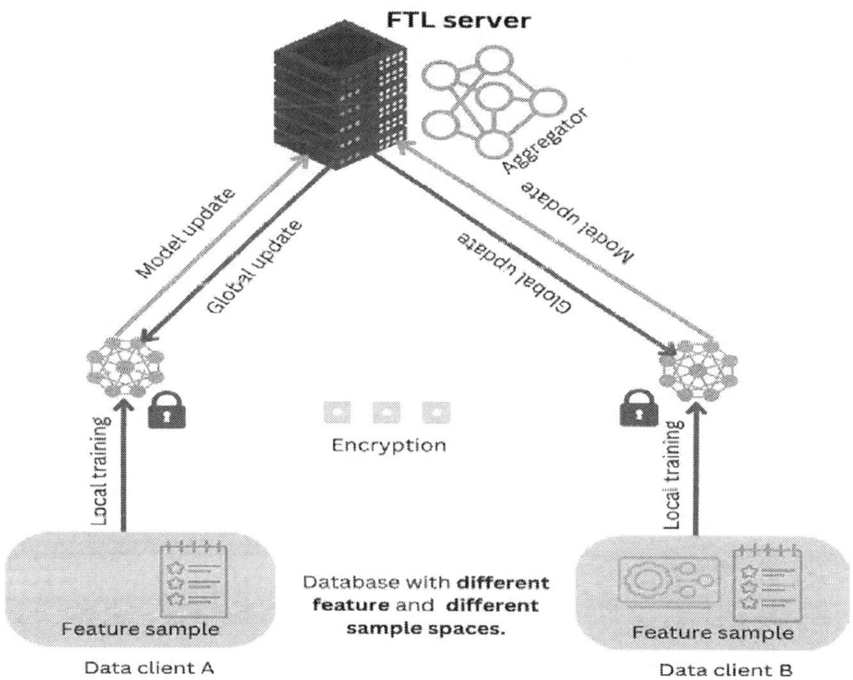


*Figure 9.6* Federated transfer learning working process.

### 9.2.4 Communication and aggregator protocols of FL

As FL evolves continuously, communication and aggregator protocol design is limitless. Therefore, this chapter categorizes the three most beneficial and useful FL approaches: synchronous, asynchronous, and semi-asynchronous FL.

#### 9.2.4.1 Synchronous FL

In this method, all participating clients update their local models simultaneously and there is a single aggregator point where aggregate these updates at a central server in a combined manner. This approach requires that all clients complete their local training and send their update to the aggregator before the central server takes action to aggregate updates to the global model, confirming the global model is consistently updated using the most recent information from all clients. This new global model is downloaded and redistributed to all local clients for the next iteration of local training [26].

As this method ensures that the global model is updated consistently across all clients, it provides stable convergence of the model. However, the need to wait for all clients to finish their updates can introduce delays, specifically if some clients have slower computational resources or larger datasets. This problem, referred to as the "straggler problem," can slow down the overall training process and reduce efficiency. The best useful scenario of this method is where consistency and synchronization across clients are critical for model performance. For example, in collaborative projects involving multiple institutions with similar computational capabilities and data availability, this method can ensure that each client contributes equally to the global model, thereby preventing any single client from disproportionately influencing the model's updates [27]. Figure 9.7 illustrates the working process of synchronous FL communication protocol.

#### 9.2.4.2 Asynchronous FL

In this method of communication, clients can update the global model independently and at different times, without requiring all clients to synchronize their updates. When clients are done with their local training, this protocol lets them send their updates to the server. The central server then adds these changes right away to the global model and makes changes to the global model. This process is continuous as new updates arrive [27]. This method addresses the "straggler problem" that is faced in the synchronous methods. This method is suitable particularly in environments where clients have varying computational processing power, network conditions, or data volumes. This method is suitable for large-scale federated learning scenarios. For example, the most useful scenario is IoT devices, mobile phones, or distributed edge computing as here heterogeneity in client capabilities is common [26]. Figure 9.8 illustrates the working process of asynchronous FL communication protocol.

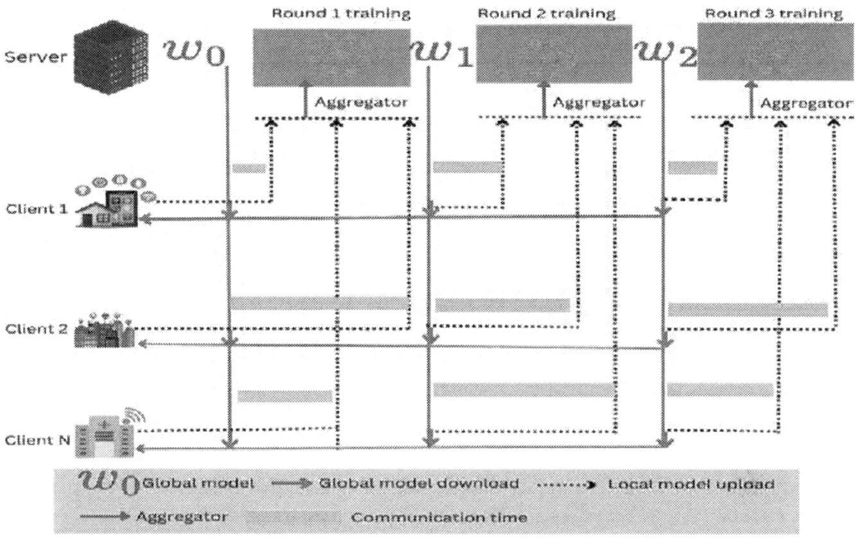


*Figure 9.7* Synchronous FL protocol working process.

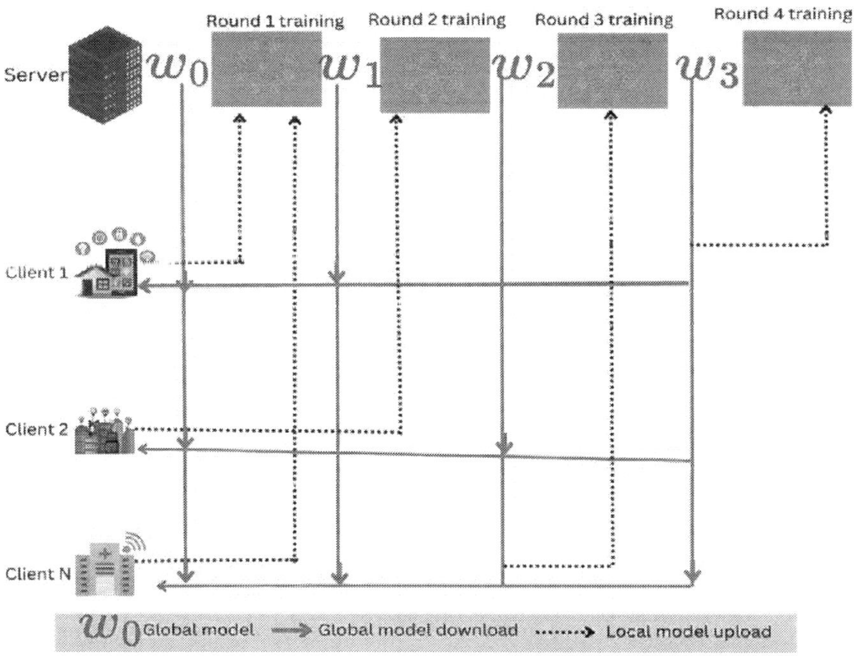


*Figure 9.8* Asynchronous FL protocol working process.

### *9.2.4.3 Semi-asynchronous FL*

This is a hybrid method as this method is the combination of both synchronous and asynchronous FL. This method permits the central server to collect updates from clients in a partially synchronized manner, while also allowing clients to publish their model updates at varied intervals. Here, the central server waits until a defined subset of clients update, or a certain threshold of updates is available before performing the aggregation in the central model [27, 28]. This approach helps balance the trade-off between efficiency and consistency by reducing the impact of stragglers without fully abandoning synchronization. In addition, this approach majorly solves the problem where delays caused by slower clients in a fully synchronous system could occur while avoiding the potential instability and model divergence that can occur in a purely asynchronous setting. In this approach, the server might be configured to aggregate updates after receiving contributions from a predetermined proportion of clients, such as 70% or 80%. This approach is well-suited for applications in mobile networks, distributed sensor networks, or any scenario where there is a mix of fast and slow clients [29]. Figure 9.9 illustrates the working process of the semi-asynchronous FL communication protocol.

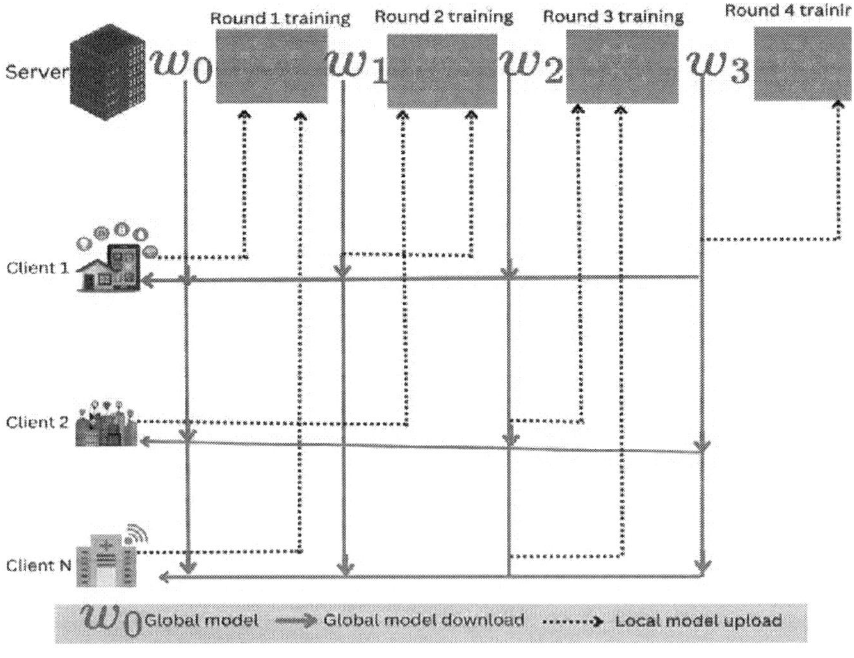


*Figure 9.9* Semi-asynchronous FL protocol working process.

## 9.3 Integration of FL with edge intelligence in IoT

The combination of FL with edge intelligence in the IoT environment is a notable progress in executing distributed ML. This method allows intelligent data processing at the edge network node. This method preserves data privacy and minimizes transmission overhead [12]. Currently, IoT devices are used almost in every sector, for instance, healthcare, industrial automation, smart homes, and smart cities. Every day these devices produce a gigantic amount of data [2]. Indeed, the need for effective and secure data processing has become crucial. This framework addresses this requirement by utilizing the processing capabilities of edge devices to train ML models collectively, hence minimizing the necessity of transmitting all of the information to a centralized server.

### *9.3.1 The role of edge intelligence in IoT*

The emergence of edge intelligence has changed the IoT environment to process IoT data. This allows the deployment of AI and ML algorithms directly at the edge network [10]. This approach reduces latency, improves real-time processing, and reduces bandwidth consumption; thus, making it as a potential component for IoT applications [7]. Also, it reduces the complete reliance on centralized cloud resources. Additionally, integrating edge intelligence with FL for IoT addresses privacy concerns through distributed model training without centralizing the data [30, 31]. This approach plays a significant role for the following reasons:

- **Minimize Latency Problem:** Edge intelligence processes data in an edge server, which minimizes data transmission to centralized servers and solves the latency problem. It provides real-time responses, which is required for most IoT applications [10].
- **Minimize Bandwidth Cost:** Edge intelligence minimizes data transmission to a central server as it runs a local model as an edge network. This saves bandwidth and lowers costs in environments with limited connectivity or high data volumes [10, 11].
- **Privacy and Security Improvement:** Local data processing through edge intelligence mitigates the risks of transmitting sensitive information, thus improving privacy and decreasing the probability of data breaches [32].
- **Improve Reliability:** Edge intelligence ensures continuous operation of devices even in the absence of a stable network connection, thus providing reliable functionality in remote or challenging environments [3].
- **Solve Storage Problem:** As edge intelligence deploys ML as the edge server and the results only go to the central server, this approach saves a considerable amount of cloud storage [9].

### 9.3.2 *Federated learning: a privacy-preserving framework*

Edge intelligence combined with FL in IoT systems establishes a robust framework for distributed ML. It prevents data privacy and uses the computational capability of edge devices. This approach handles the challenges of IoT environments. Some challenges are data heterogeneity, bandwidth constraints, latency problems, and the need for real-time data processing [6, 12]. The framework usually contains multiple layers where each layer serves crucial functions contributing to the system.

#### *9.3.2.1 Architecture layers*

i. IoT layer
   a. Components: This layer consists of various IoT devices. These constantly generate data from their environment [3].
   b. Workflow: The IoT layer generates and sends data to the edge intelligence layer for further processing [4, 33].

ii. Edge intelligence layer
   a. Components: This layer includes edge devices such as gateways, routers, Raspberry Pi, Arduino, or edge servers that are positioned closer to the data sources [12].
   b. Workflow: This layer performs complex data processing, which includes feature extraction and local model training. Edge devices in this layer train local models using the data collected from the IoT layer [12].
   c. Architecture elements
      i. Model training: Edge devices train ML models locally using the local data.
      ii. Communication protocol: Communication between IoT end device and edge intelligence layers can follow synchronous, asynchronous, or semi-asynchronous protocols. The protocol choice impacts the timing and coordination of data processing tasks and the transmission of model updates [26].
      iii. Model update aggregation: Aggregation can occur at the edge device, or local model results can be passed on to the central server or other peer devices in a P2P network architecture based on the application scenario [27].

iii. Centralized server or distributed network architectures
   a. Components: There are two types of architecture. A centralized server (cloud server) or a P2P network FL (distributed server). The most popular one is the centralized server architecture [24]. Figure 9.10 illustrates the details of a centralized FL edge intelligence IoT architecture.

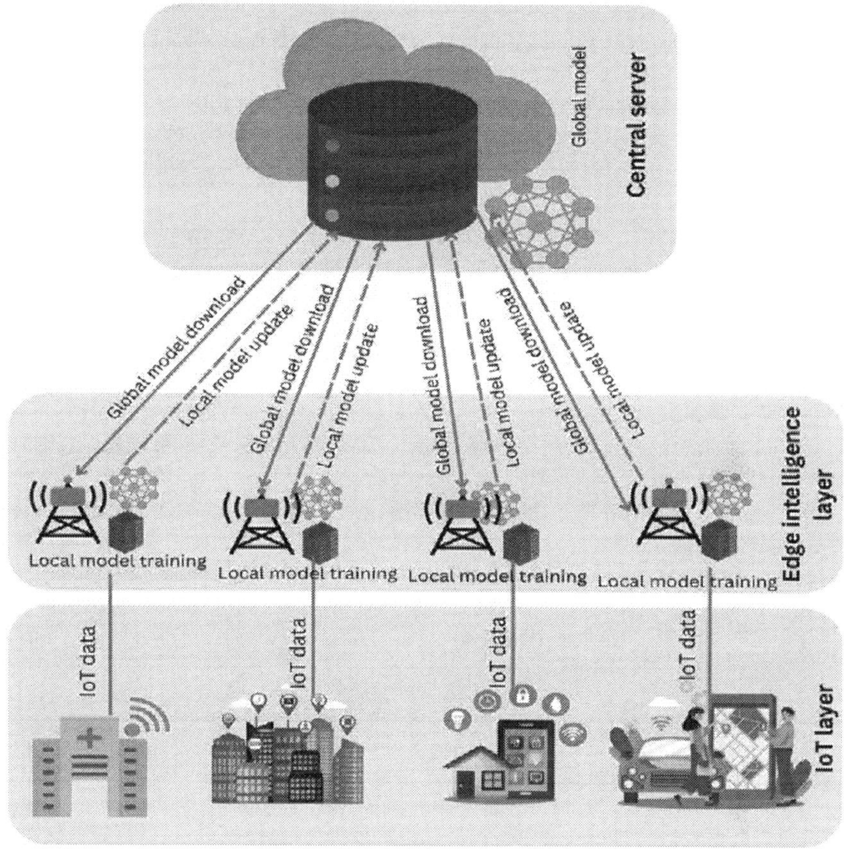


*Figure 9.10* FL edge intelligence IoT architecture layer working process.

b. Workflow: In a centralized architecture, the central server aggregates model updates from all participating edge devices, applies secure aggregation techniques, and updates the global model. In a P2P architecture, model updates are shared and aggregated directly between devices where each device contributes to the final model [25].
c. Architecture elements
   i. Global model training: All the edge servers will send their local model results as a parameter or gradient to the central server to train the global model. For a distributed architecture, they send the results to another neighboring edge server to train the model [3].
   ii. Communication protocol: Communication between the edge server and cloud server can follow synchronous, asynchronous,

or semi-asynchronous protocols. The protocol choice impacts the timing and coordination of data processing tasks and the transmission of model updates [26].

iii. Global model distribution: The aggregated global model is then redistributed to the edge server. This iterative process continues until an optimal state [27].

### *9.3.3 Applications of FL-based edge intelligence in IoT*

#### *9.3.3.1 Smart healthcare: predictive health monitoring*

**Use case scenario:** A network of wearable health devices that includes smartwatches and fitness trackers. It can monitor chronic diseases such as diabetes and hypertension across a population. These devices continuously collect data; for instance, heart rate, blood glucose levels, and physical activity. Each device locally trains a model on its user's health data through using the FL approach, predicting potential health issues such as an impending heart attack or an increase or reduction of blood sugar level [30].

**Solution:** Each device (edge server) trains a local model using their own data (IoT device). This model updates then sends the result to a central server or P2P network to aggregate these updates to make a global predictive health model. Healthcare providers can offer personalized care recommendations based on collective network intelligence where sensitive patient data remain on the user's device. This FL-based edge intelligence framework is particularly effective in remote health monitoring and management systems, where patient privacy is crucial [3].

#### *9.3.3.2 Autonomous vehicles: collaborative driving intelligence*

**Use case scenario:** Autonomous vehicles continually gather sensor data. Some examples of sensor data are those from cameras, Light Detection and Ranging, and Global Positioning System, to navigate and make driving decisions. Each vehicle has its own driving conditions based on weather, traffic patterns, and road types. These vehicles learn from their experiences and contribute to a shared driving model through the FL technique [34].

**Solution:** Each autonomous vehicle (edge server) trains its driving algorithms using its own collected data. Then it shares model updates to a central server or directly to other closer vehicles in a P2P network. These updates aggregate to make a global driving model that helps to improve safety and efficiency across all participating vehicles. This approach preserves privacy of the sensitive data, for example, location and driving habits do not need to be sent to the central server. This FL-based edge intelligence is useful in collaborative autonomous driving systems, where real-time data sharing is crucial for safety and performance [2, 3, 35].

#### *9.3.3.3 Smart cities: traffic management optimization*

**Use case scenario:** In a smart city, there are a lot of IoT devices, including traffic cameras, environmental sensors, and smart streetlights, which collect data to manage urban infrastructure. For example, traffic cameras monitor vehicle flow at intersections, while environmental sensors track air quality. The city optimizes traffic management to reduce congestion and improve air quality [36].

**Solution:** These devices (edge server) train local models using local data to predict traffic congestion or pollution levels. The devices then share their updates with a central server or a distributed network to create a global model. Through this collaborative model training, the city can optimize traffic light timings, reduce vehicle waiting, and improve overall traffic flow. This approach helps urban traffic management without centralizing sensitive data, which ensures residential privacy and data security. This FL-based edge intelligence is useful in large-scale urban traffic management systems, where confidentiality and real-time data processing are required [36, 37].

## 9.4 Challenges and future directions

As discussed in the previous sections, edge intelligence can improve the IoT applications through implementing the FL technique. Although a thorough survey highlights the considerable promise and benefits, it also identifies some research problems that need to be addressed to adopt FL-based edge intelligence for IoT systems in the future.

- The current focus of research on technologies address some challenges; for example, resource constraints, data and system heterogeneity, where privacy and security in edge computing are extensive. However, there is a need to pay more attention to the dynamic characteristics of large-scale edge computing systems for IoT applications [7]. These dynamics affect the performance of FL in edge environments. Much of the existing FL research assumes a stable and static system where FL servers are consistently available, network connections are stable, and client devices have adequate computational resources [14, 38]. However, these assumptions must be revised at a large-scale and scholars can give attention to the dynamic edge environments. For example, wireless communication states in sixth generation (6G) networks can fluctuate within milliseconds [3], which is far shorter than typical FL training periods, posing a challenge to the stability of FL systems.
- Current security solutions for FL and edge computing for IoT environments require more computational resources than these devices can offer [9]. Their integration still needs to be explored. Some research has already

existed for optimization [39, 40]. Future research must prioritize the development of lightweight, interoperable security solutions where the focus will be optimization and protection of privacy across heterogenous and distributed edge environments.

- The majority of the FL research has been done to solve privacy issues, especially parameter or weight transfer to a central server (to reduce data leakage/data breaches.) Privacy issues in edge FL have not been well studied as just a few recent works focus on them [7]. The privacy in edge computing can be an issue as malicious users may access sensitive information (e.g., data and usage patterns, location) of an edge node that they are not allowed to access [41]. This research direction adds more consideration for edge computing as it is considered a potential new path to focus on for the preservation of privacy in edge federated learning.
- There is a lot of literature work available on training non-IID data using asynchronous and semi-asynchronous FL in distributed and decentralized IoT environments. The focus is mostly on the privacy aspect of data as an alternative solution for a centralized ML approach [27, 36, 42, 43]. However, a major gap exists in finding optimized worker nodes in the FL network. This optimization will be crucial for improving the efficiency of the learning process during communication rounds, particularly with the prevalent non-IID data in the IoT ecosystem.

## 9.5 Conclusion

In this chapter, we studied the significance of FL and edge intelligence within the IoT context. We began by discussing the fundamental aspects of FL. It includes its core characteristics, architecture, category, and the various communication and aggregation protocols. We then examined how FL integrates with edge intelligence to improve IoT systems. We discussed the privacy-preserving FL-based edge intelligence architecture. We highlighted several practical applications of FL-driven edge intelligence in IoT application scenarios where we illustrated its potential to transform data processing and decision-making in distributed environments. Finally, we identified key challenges that need to study in this field, along with possible future research directions. This chapter discussed the critical role of FL in advancing edge intelligence for IoT. It offers a secure and efficient framework that handles complex and resource-constrained IoT ecosystems through edge intelligence.

**Acknowledgment:** This work was supported by the Research Council of Finland with CHIST-ERA, Grant agreement no—359790, Di4SPDS-Distributed Intelligence for Enhancing Security and Privacy of Decentralized and Distributed Systems.

## References


1. Shanmugam, Lavanya, Ravish Tillu, and Manish Tomar. "Federated learning architecture: Design, implementation, and challenges in distributed AI systems." Journal of Knowledge Learning and Science Technology 2, no. 2 (2023): 371–384.
2. Nguyen, Dinh C., Ming Ding, Pubudu N. Pathirana, Aruna Seneviratne, Jun Li, and H. Vincent Poor. "Federated learning for internet of things: A comprehensive survey." IEEE Communications Surveys & Tutorials 23, no. 3 (2021): 1622–1658.
3. Duan, Qiang, Jun Huang, Shijing Hu, Ruijun Deng, Zhihui Lu, and Shui Yu. "Combining federated learning and edge computing toward ubiquitous intelligence in 6G network: Challenges, recent advances, and future directions." IEEE Communications Surveys & Tutorials 25 (2023): 2892–2950.
4. Kotecha, Prajay, Tanvi Dhoka, Jitendra Bhatia, Malaram Kumhar, Rajesh Gupta, Sudeep Tanwar, and Nilesh Kumar Jadav. "Performance evaluation of federated learning in edge computing environment." Procedia Computer Science 235 (2024): 2955–2964.
5. Amin, Syed Umar, and M. Shamim Hossain. "Edge intelligence and Internet of Things in healthcare: A survey." IEEE Access 9 (2020): 45–59.
6. Lim, Wei Yang Bryan, Nguyen Cong Luong, Dinh Thai Hoang, Yutao Jiao, Ying-Chang Liang, Qiang Yang, Dusit Niyato, and Chunyan Miao. "Federated learning in mobile edge networks: A comprehensive survey." IEEE Communications Surveys & Tutorials 22, no. 3 (2020): 2031–2063.
7. Xia, Qi, Winson Ye, Zeyi Tao, Jindi Wu, and Qun Li. "A survey of federated learning for edge computing: Research problems and solutions." High-Confidence Computing 1, no. 1 (2021): 100008.
8. Lim, W. Y. B., Luong, N. C., Hoang, D. T., Jiao, Y., Liang, Y. C., Yang, Q., Niyato, D. and Miao, C., 2020. Federated learning in mobile edge networks: A comprehensive survey. IEEE Communications Surveys & Tutorials 22, no. 3: 2031–2063.
9. Salh, Adeb, Razali Ngah, Lukman Audah, Kwang Soon Kim, Qazwan Abdullah, Yahya M. Al-Moliki, Khaled A. Aljaloud, and Hairul Nizam Talib. "Energy-efficient federated learning with resource allocation for green IoT edge intelligence in B5G." IEEE Access 11 (2023): 16353–16367.
10. Wu, Qiong, Kaiwen He, and Xu Chen. "Personalized federated learning for intelligent IoT applications: A cloud-edge based framework." IEEE Open Journal of the Computer Society 1 (2020): 35–44.
11. Ramya, R., and S. Ramamoorthy. "Survey on edge intelligence in IoT based computing platform." Ambient Communications and Computer Systems: Proceedings of RACCCS 2021 (2022): 549–561.
12. Zhao, Lei, Lin Cai, and Wu-Sheng Lu. "Transform-domain federated learning for edge-enabled IoT intelligence." IEEE Internet of Things Journal 10, no. 7 (2022): 6205–6220.
13. Konečný, Jakub. "Federated learning. Strategies for improving communication efficiency." arXiv preprint arXiv:1610.05492 (2016).
14. Khan, Latif U., Walid Saad, Zhu Han, Ekram Hossain, and Choong Seon Hong. "Federated learning for internet of things: Recent advances, taxonomy, and open challenges." IEEE Communications Surveys & Tutorials 23, no. 3 (2021): 1759–1799.
15. Luzón, M. Victoria, Nuria Rodríguez-Barroso, Alberto Argente-Garrido, Daniel Jiménez-López, Jose M. Moyano, Javier Del Ser, Weiping Ding, and Francisco

Herrera. "A tutorial on federated learning from theory to practice: Foundations, software frameworks, exemplary use cases, and selected trends." IEEE/CAA Journal of Automatica Sinica 11, no. 4 (2024): 824–850.
16. Li, Li, Yuxi Fan, Mike Tse, and Kuo-Yi Lin. "A review of applications in federated learning." Computers & Industrial Engineering 149 (2020): 106854.
17. Li, Tian, Anit Kumar Sahu, Ameet Talwalkar, and Virginia Smith. "Federated learning: Challenges, methods, and future directions." IEEE Signal Processing Magazine 37, no. 3 (2020): 50–60.
18. Pfeiffer, Kilian, Martin Rapp, Ramin Khalili, and Jörg Henkel. "Federated learning for computationally constrained heterogeneous devices: A survey." ACM Computing Surveys 55, no. 14s (2023): 1–27.
19. Lim, Wei Yang Bryan, Nguyen Cong Luong, Dinh Thai Hoang, Yutao Jiao, Ying-Chang Liang, Qiang Yang, Dusit Niyato, and Chunyan Miao. "Federated learning in mobile edge networks: A comprehensive survey." IEEE Communications Surveys & Tutorials 22, no. 3 (2020): 2031–2063.
20. Uddin, Md Raihan, Gauri Shankar, Saddam Hossain Mukta, Prabhat Kumar, and Najmul Islam. "Evolving topics in federated learning: Trends, and emerging directions for IS." arXiv preprint arXiv:2409.15773 (2024).
21. Zhang, Chen, Yu Xie, Hang Bai, Bin Yu, Weihong Li, and Yuan Gao. "A survey on federated learning." Knowledge-Based Systems 216 (2021): 106775.
22. Islam, Mohammad Munzurul, and Mohammed Alawad. "Optimizing Federated Learning with Heterogeneous Edge Devices." In 2024 IEEE 3rd International Conference on Computing and Machine Intelligence (ICMI), pp. 1–5. IEEE, 2024.
23. Shanmugam, Lavanya, Ravish Tillu, and Manish Tomar. "Federated learning architecture: Design, implementation, and challenges in distributed AI systems." Journal of Knowledge Learning and Science Technology 2, no. 2 (2023): 371–384.
24. Narula, Manu, Jasraj Meena, and Dinesh Kumar Vishwakarma. "A comprehensive review on federated learning for data-sensitive application: Open issues & challenges." Engineering Applications of Artificial Intelligence 133 (2024): 108128.
25. Ge, Lina, Haiao Li, Xiao Wang, and Zhe Wang. "A review of secure federated learning: Privacy leakage threats, protection technologies, challenges and future directions." Neurocomputing (2023): 126897.
26. Qi, Pian, Diletta Chiaro, Antonella Guzzo, Michele Ianni, Giancarlo Fortino, and Francesco Piccialli. "Model aggregation techniques in federated learning: A comprehensive survey." Future Generation Computer Systems 150 (2024): 272–293.
27. Xu, Chenhao, Youyang Qu, Yong Xiang, and Longxiang Gao. "Asynchronous federated learning on heterogeneous devices: A survey." Computer Science Review 50 (2023): 100595.
28. Wu, Wentai, Ligang He, Weiwei Lin, Rui Mao, Carsten Maple, and Stephen Jarvis. "SAFA: A semi-asynchronous protocol for fast federated learning with low overhead." IEEE Transactions on Computers 70, no. 5 (2020): 655–668.
29. Zhang, Jinfeng, Wei Liu, Yejun He, Zhou He, and Mohsen Guizani. "Semi-asynchronous model design for federated learning in mobile edge networks." IEEE Transactions on Vehicular Technology 72 (2023): 16280–16292.
30. Akter, Mahmuda, Nour Moustafa, Timothy Lynar, and Imran Razzak. "Edge intelligence: Federated learning-based privacy protection framework for smart healthcare systems." IEEE Journal of Biomedical and Health Informatics 26, no. 12 (2022): 5805–5816.

31. Smmarwar, Santosh K., Govind P. Gupta, and Sanjay Kumar. "AI-empowered malware detection system for industrial internet of things." Computers and Electrical Engineering 108 (2023): 108731.
32. Lv, Lingling, Zongyu Wu, Lei Zhang, Brij B. Gupta, and Zhihong Tian. "An edge-AI based forecasting approach for improving smart microgrid efficiency." IEEE Transactions on Industrial Informatics 18, no. 11 (2022): 7946–7954.
33. Kumar, Prabhat, Govind P. Gupta, and Rakesh Tripathi. "PEFL: Deep privacy-encoding-based federated learning framework for smart agriculture." IEEE Micro 42, no. 1 (2021): 33–40.
34. Chellapandi, Vishnu Pandi, Liangqi Yuan, Christopher G. Brinton, Stanislaw H. Żak, and Ziran Wang. "Federated learning for connected and automated vehicles: A survey of existing approaches and challenges." IEEE Transactions on Intelligent Vehicles 9 (2023): 119–137.
35. Anjos, Julio C. S. Dos, Kassiano J. Matteussi, Fernanda C. Orlandi, Jorge L. V. Barbosa, Jorge Sá Silva, Luiz F. Bittencourt, and Cláudio F. R. Geyer. "A survey on collaborative learning for intelligent autonomous systems." ACM Computing Surveys 56, no. 4 (2023): 1–37.
36. Pandya, Sharnil, Gautam Srivastava, Rutvij Jhaveri, M. Rajasekhara Babu, Sweta Bhattacharya, Praveen Kumar Reddy Maddikunta, Spyridon Mastorakis, Md Jalil Piran, and Thippa Reddy Gadekallu. "Federated learning for smart cities: A comprehensive survey." Sustainable Energy Technologies and Assessments 55 (2023): 102987.
37. Zheng, Zhaohua, Yize Zhou, Yilong Sun, Zhang Wang, Boyi Liu, and Keqiu Li. "Applications of federated learning in smart cities: Recent advances, taxonomy, and open challenges." Connection Science 34, no. 1 (2022): 1–28.
38. Liu, Yaqiong, Mugen Peng, Guochu Shou, Yudong Chen, and Siyu Chen. "Toward edge intelligence: Multiaccess edge computing for 5G and Internet of Things." IEEE Internet of Things Journal 7, no. 8 (2020): 6722–6747.
39. Mills, Jed, Jia Hu, and Geyong Min. "Communication-efficient federated learning for wireless edge intelligence in IoT." IEEE Internet of Things Journal 7, no. 7 (2019): 5986–5994.
40. Zhang, Tinghao, Kwok-Yan Lam, and Jun Zhao. "Device scheduling and assignment in hierarchical federated learning for Internet of Things." IEEE Internet of Things Journal 11 (2024): 18449–18462.
41. Li, Haiao, Lina Ge, and Lei Tian. "Survey: Federated learning data security and privacy-preserving in edge Internet of Things." Artificial Intelligence Review 57, no. 5 (2024): 130.
42. Yu, Dongxiao, Zhenzhen Xie, Yuan Yuan, Shuzhen Chen, Jing Qiao, Yangyang Wang, Yong Yu, Yifei Zou, and Xiao Zhang. "Trustworthy decentralized collaborative learning for edge intelligence: A survey." High-Confidence Computing (2023): 100150.
43. Chen, Qimei, Zehua You, and Hao Jiang. "Semi-asynchronous hierarchical federated learning for cooperative intelligent transportation systems." arXiv preprint arXiv:2110.09073 (2021).

Chapter 10

# Deep learning–based edge intelligence for IoT system

*Neerav Pawar, Md Raihan Uddin, and Gauri Shankar*

## 10.1 Introduction

Convergence of edge computing with deep learning (DL) addresses certain concerns regarding (i) delayed responses and (ii) security and privacy breaches of sensitive private data by performing the training and inference at the edge itself, which is near the data source. As in the network, many devices are connected, and traditional cloud-centric models struggle to keep up with the demands for low latency, high bandwidth efficiency, enhanced privacy, and reliable real-time processing. Cloud computing meets the resource demands due to its exceptional computing power, which is a requirement for training a DL model [1]. This chapter explores the relationship between edge computing and DL, exploring how their integration can uncover the full potential of Internet of Things (IoT) systems. Edge computing is a technique in intelligent IoT systems that enables users to execute computation on the edge devices connected to the network. This is performed on downstream data in place of cloud services and upstream data in place of IoT devices [2]. This distributed computing paradigm processes data near the data collection source rather than being processed by the central data processing node. This approach significantly reduces latency, optimizes bandwidth usage, and enhances data privacy and security. For devices connected at the edge of the network, edge computing enables them to process tasks at the edge itself; otherwise, they will be offloaded to the cloud server, facilitating efficient and quick data handling. DL techniques are useful in enabling smart edge computing by providing multiple layers of computing neural networks in a machine learning mode. DL allows networks to make decisions to analyze big data as it is capable of learning. Models such as Convolutional Neural Networks (CNNs), Graph Neural Networks (GNNs), and Recurrent Neural Networks (RNNs) have applications in image recognition, natural language processing (NLP), and other domains. IoT systems could process and execute data locally by using DL algorithms [3].

IoT devices connected to the network are directly affected by the energy that is being consumed. When offloading tasks, for instance, if the task is processed and executed by the edge device itself, it consumes its own energy, but

DOI: 10.1201/9781003512066-10

when the task is offloaded to the server for processing, it calls for extra energy consumption resulting in communication latency and execution delay [4].

The ability to process data locally using DL algorithms is important for applications that require immediate actions and insights. For instance, in smart homes, real-time monitoring and anomaly detection are necessary for security and automation. In healthcare, wearable devices that are equipped with DL capabilities can provide continuous patient monitoring and early diagnostics. Industrial IoT (IIoT) applications benefit from predictive maintenance and quality control; automated alerts generated by edge devices in the event of disruption or detection could eliminate the response time significantly and will allow the risk of mitigation at an early stage [5]. Edge computing architecture involves a layered approach where data processing is distributed across edge devices, nodes, and the central cloud. This architecture ensures that data can be processed quickly and efficiently, with minimal reliance on the cloud, reducing latency and bandwidth usage. Different neural networking algorithms, such as CNNs, GNNs, and RNNs, are optimized to deploy on edge devices. Approaches, such as transfer learning or TinyML (machine learning on microcontrollers) allow the adaptation of complex models to run efficiently on resource-constrained devices. The execution of DL models on edge devices is facilitated by specialized hardware such as Graphics Processing Units (GPUs), Tensor Processing Units (TPUs), and dedicated artificial intelligence (AI) chips like NVIDIA Jetson and Google Coral. These hardware components are designed to accelerate neural network computations, making running DL algorithms locally feasible, due to the following reasons.

Reduced Latency: Processing data at the edge significantly reduces the time taken for data to travel to and from the central server, ensuring faster response times. When data are handled locally, bandwidth is minimized, minimizing the amount of data transferred to the cloud. Thus, edge computing minimizes the data transmitted to the cloud, conserving bandwidth [5].

The chapter further explores practical applications of edge-based DL models in IoT domains, such as healthcare, smart homes, smart cities, industrial IoT, and agriculture. Each application area is discussed with specific examples and use cases demonstrating the benefits and effectiveness of integrating DL with edge computing [6].

Enhanced Privacy and Security: Local data processing ensures that sensitive information can be analyzed and acted upon without leaving the device, protecting the data from any possible data breaches during transmission.

Energy Efficiency: Optimized DL models and specialized hardware contribute to lower energy consumption, which is critical for battery-powered IoT devices.

Scalability and Flexibility: Edge-based DL systems can scale more effectively as they do not rely solely on cloud resources and can be adapted to various use cases and environments.

Reliability and Resilience: Local processing capabilities provide fallback options in case of network failures, ensuring continuous operation of critical applications.

Cost Efficiency: Reducing the dependency on cloud resources lowers operational costs, making IoT deployments more economical.

Personalization and Context-Awareness: Edge devices can use real-time data to provide personalized and context-aware services, enhancing user experience and functionality. Hence, decisions can be made instantaneously [7].

Despite its numerous advantages, implementing DL on edge devices presents several challenges. These include resource constraints, model optimization techniques, data management, and the deployment and maintenance of edge systems. The chapter addresses these challenges and discusses potential solutions. The chapter concludes with detailed case studies and experimental results, comparing edge-based DL applications with traditional cloud-based approaches. Key performance metrics such as latency, energy consumption, accuracy, and scalability are evaluated to highlight the efficacy of edge intelligence in IoT systems [9]. The chapter identifies future trends and research directions in edge AI, including advances in edge AI hardware, emerging DL algorithms, integration with fifth generation (5G) networks and beyond, edge federated learning, autonomous and self-learning edge systems, and ethical and regulatory considerations. The integration of DL with edge computing is poised to revolutionize IoT systems, offering substantial improvements in performance, efficiency, and functionality [10].

We seek to answer the following research questions:

1. What components should a DL-based edge intelligent IoT system have?
2. What kind of deep learning models can be deployed on edge devices?
3. What might be possible challenges to deploying such a model on edge devices, and how can they be addressed?
4. What potential do intelligent IoT systems hold for future and how DL could accelerate their adoption in various industries?
5. What advantages do edge-based systems have over cloud-based systems?

Through these research questions, this chapter aims to give a comprehensive overview of the state-of-the-art in DL-based edge intelligence, highlighting the advantages, challenges, and future directions for innovation. This discussion underscores the transformative potential of combining DL with edge intelligence in IoT systems, offering a roadmap for future research and development in this rapidly evolving domain. Section 2 describes the proposed DL-based architecture. It discusses hardware, software requirements, and types DL-based models that can be used to make edge devices intelligent. Section 3 of this chapter highlights the advantages of integrating these DL techniques to make IoT systems more efficient. Section 4 discusses the real-world application of intelligent edge-based models and their future scope in various domains. Section 5 highlights possible challenges that could arise while deploying DL models on edge devices in IoT systems. Section 6 answers the fourth question posed earlier, which presents the experimental case studies that highlight the possible role of intelligent edge-based IoT systems in enhancing day-to-day activities in various industries.

## 10.2 Architecture and design of edge-based DL systems

The combination of deep learning (DL) models into IoT systems that rely on edge computing has proven to be effective and practical. This enables the user to have quick results due to real-time data processing and intelligent decision-making right where the data are generated. This section talks about how DL-based intelligent edge computing IoT systems are designed and built, the model architecture, components, software, and hardware used [11].

Figure 10.2 showcases a distributed and hierarchical architecture of DL-based intelligent edge computing systems. Edge devices generate data, which are further processed at the edge of things (EoT). The edge of things is the edge computing layer in the architecture displayed in Figure 10.1. EoT processes the generated data from edge devices before transmitting it to other nodes or to the cloud, enabling it to handle the data closer to its generation source. The data are generated by sensors to edge nodes, where initial

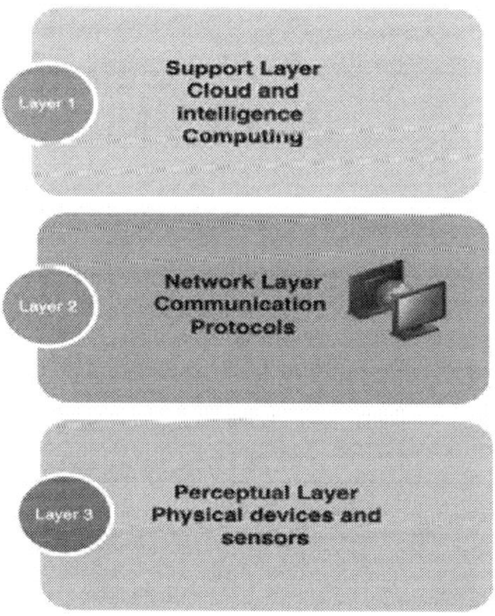


*Figure 10.1* A generic IoT network security architecture [8].

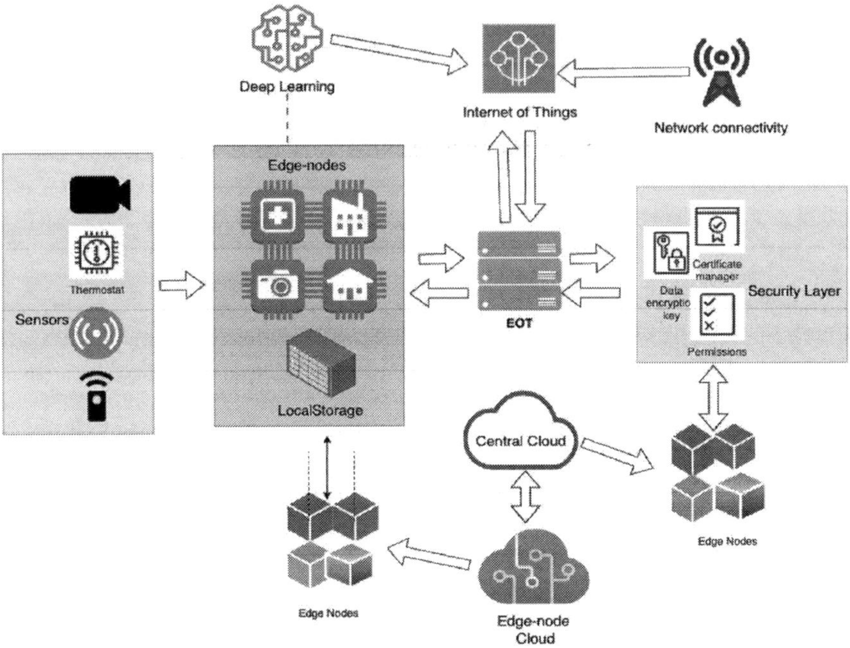


*Figure 10.2* Architecture and design of edge-based DL cloud-enabled IoT systems.

processing is done. The processed data are stored locally and then sent to the EOT for distribution over the system, or it is securely sent to the cloud for any additional processing or long-term storage.

This enables the data transferring capability of the model to go faster as data processing and transmission speeds up. The architecture usually has the following three main parts:

- *Edge Devices*: These gadgets and sensors are responsible for gathering data. For example, smartphones, thermostats, or a security camera. They collect data and handle basic processing tasks, but they have limited computing power.
- *Edge Nodes/Gateways*: They are powerful devices that collect data from edge devices. Edge nodes are responsible for complex processing to make quick decisions without always having to send the data to the cloud. They act like local decision-making managers for the cloud.
- *Security layer*: It ensures the data are only accessible to authorized entities, making them secure throughout the process.
- *Sensors*: Sensors like cameras, thermostats, and others gather data from the surrounding environment, like temperature readings, video footage, humidity or moisture readings, or other sensor specific metrics.

- *Cloud Servers*: The cloud servers handle the most complex and heavy processing tasks. They are also responsible for long-term storage. This is due to their computing power and capability to store significant amounts of data.

### *10.2.1 DL model architectures for edge devices*

DL models are run on edge devices connected to the IoT network, as defined in Table 10.1. This can be tricky sometimes, as these devices may have limited processing power or memory constraints. However, applying smart design and optimization techniques can help. The following are some models that have proven to have applications in such environments:

- Convolutional Neural Networks (CNNs) have proven to be coherent for analyzing images and patterns. They could pick details and patterns of data points, which make them ideal for tasks such as identifying objects or facial recognition from photographs [11].
- Graph Neural Networks (GNNs) handle data which present themselves in the form of a network, for example, a social network or a transport network. In this web of information, GNNs excel in discovering patterns and connections [12].

*Table 10.1* Model architecture comparison based on latency, bandwidth, and inference time.

| *Model Architecture* | *Latency* | *Bandwidth Usage* | *Inference Time* |
|---|---|---|---|
| Convolutional Neural Networks (CNNs) | Moderate Latency due to convolution operations | Moderate to High (depends on image size) | Moderate (depends on the depth of the network and image size) |
| Graph Neural Networks (GNNs) | High Latency (due to complex graph operations) | Low to Moderate (data are often sparse) | High (depends on the complexity of the graph structure) |
| Recurrent Neural Networks (RNNs) | Low Latency for short sequences, higher for long sequences | Low to Moderate (based on sequence length) | Moderate to High (depends on sequence length) |
| Transfer Learning | Low Latency (after initial model loading) | Very Low (designed for minimal data transfer) | Very Low (optimized for fast inference on microcontrollers) |
| TinyML | Very Low Latency (optimized for small devices) | Very Low (designed for minimal data transfer) | Very Low (optimized for fast inference on microcontrollers) |

- Recurrent Neural Networks (RNNs) perfectly handle sequential data like series of numbers or texts. These neural networks are great at picking up patterns that tend to unfold over time, hence making them handy for tasks such as prediction models or speech recognition models [13].
- Transfer learning utilizes a pretrained model and fine-tunes it to execute a specific task. This model saves resources and time as the model already has learned the general features from the previous large dataset [14].
- TinyML runs machine learning models and algorithms on small devices with relatively small computing power such as microcontrollers. These models are generally optimized in such a way that they use little resources and power [15].

#### *10.2.1.1 Hardware requirements for edge artificial intelligence*

Specialized hardware is required to run these models, as follows:

I. *GPUs and TPUs*: GPUs and TPUs that can handle heavy computations are required by DL models. These units make the running of these models efficient and quick [16].
II. *Specialized Edge AI Chips*: Technology giants such as Google and NVIDIA are contributing to the IoT domain by introducing specialized chips like NVIDIA Jetson and Google Coral. These chips are specially engineered to fulfill the requirements of an AI model on the edge of an intelligent IoT system. They provide all the power without using unnecessary energy, and are supported by a range of software frameworks and tools.

The following are some examples of software tools or frameworks that help in deploying DL models on edge computing devices:

I. *TensorFlow Lite*: This is the mobile version of TensorFlow that can be run on mobile devices and embedded systems. This lightweight version runs machine learning (ML) models on edge devices efficiently.
II. *PyTorch Mobile*: Like TensorFlow Lite, Pytorch mobile deploys Pytorch models on edge and mobile devices providing all the necessary support to run the model. Pytorch mobile is easy to use and very flexible.
III. *Edge Impulse*: This is designed to develop ML models for edge devices. The platform provides all the tools to facilitate this including tools for model training, data collection, and model deployment, hence making it easier for implementing AI solutions to mobile and edge devices.

## 10.3 Advantages of DL on edge computing

There are numerous benefits for using DL on edge devices in an intelligent IoT system. Combining these could enhance efficiency, security, and overall performance, as shown in Figure 10.3.

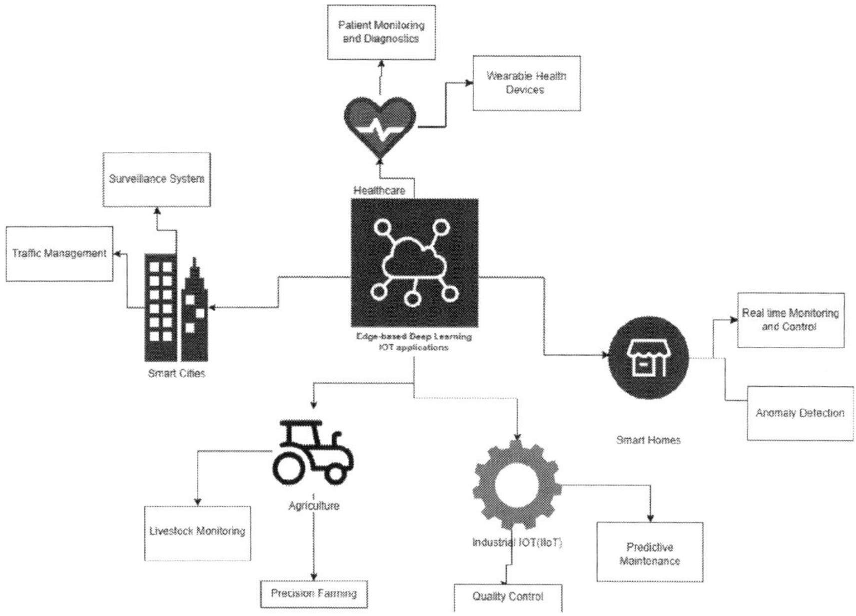


*Figure 10.3* Applications of edge-based DL IoT systems in various sectors.

### *10.3.1 Reduced latency and bandwidth efficiency*

One of the biggest advantages of processing data on the edge is reduced latency. When data are analyzed locally, there's no need to send it to a cloud server and wait for a response. This means faster decision-making, which is crucial for things like self-driving cars, industrial robots, and healthcare monitors [17]. Edge computing saves bandwidth by handling most of the data processing locally. Only important data that need further analysis or long term storage get transmitted to the cloud. This reduces the amount of data traveling within the network, which is great for places with constrained bandwidth or high data costs [18].

### *10.3.2 Enhanced privacy, security, and energy efficiency*

Processing data on the edge helps keep it secure. Sensitive information can be handled locally, reducing the risk of it being intercepted during transmission. This is especially important in areas like healthcare and smart homes, where personal data are often in play [19]. Edge devices are designed to be energy efficient. Running DL models on these devices helps conserve energy, which is critical for battery-powered gadgets like wearables and remote sensors. Optimized models and special hardware can perform complex tasks without using too much power [20].

### 10.3.3 Scalability, flexibility, and improved performance

Edge systems can easily scale as you add more devices. Unlike cloud systems that might need major upgrades to handle more data, edge devices can process data independently. This makes it easy to scale up and tailor devices to specific tasks and environments without overloading a central system [19, 20]. Local data processing leads to better performance of IoT applications. Decisions are made instantly, and actions are taken without the delay of cloud processing. This is crucial for apps needing immediate feedback, like interactive gaming, augmented reality, and real-time control systems [21].

### 10.3.4 Reliability, resilience, and cost-effectiveness

Local processing makes IoT systems more reliable. Edge devices can keep working and making decisions even if they lose connection to the cloud. This is vital for critical applications like industrial automation and emergency services, where downtime isn't an option. Edge computing reduces reliance on cloud resources, which can lower costs. With less data needing to be sent back and forth to the cloud, data transfer fees drop, and the need for extensive cloud infrastructure diminishes. This makes edge computing a cost-effective choice for large-scale IoT projects [22].

### 10.3.5 Personalization and support for emerging applications

Edge devices can offer personalized services by analyzing real-time data specific to each user. In smart homes, devices can learn your preferences and adjust settings automatically. In retail, edge devices can enhance your shopping experience by responding to your behavior and preferences in real time [23]. Edge-based DL enables new and innovative applications that need real-time data processing. Think of smart cities, where traffic systems must adapt on the fly, or precision farming, where real-time analysis of environmental data can improve crop management [21].

## 10.4 Applications of edge-based DL in IoT

Edge-based DL is revolutionizing how IoT technologies are used in various fields. Edge devices can make quick decisions and provide real-time insights by processing data locally. The following are some critical applications in different areas:

### 10.4.1 Smart homes

- *Real-time Monitoring, Control and Anomaly Detection*

In smart homes, edge-based DL allows devices to moderate and supervise home environments in real time. For example, smart thermostats can learn your heating preferences and adjust settings automatically. Security systems can instantly alert you to any unusual activity, providing peace of mind [24]. Smart home devices equipped with DL can detect anomalies. For example, if one of the windows of the house is opened unexpectedly or the system detects an unusual movement when the house is supposed to be empty, an alert is sent to the homeowners about the irregularity. This prevents thefts or break-ins into the house, ensuring the safety and security of the property [25], as shown in Figure 10.4.

### 10.4.2 Healthcare

- *Patient Monitoring, Diagnostics, and Use of Wearable Health Devices*
  Patient care is transforming due to edge-based DL. Wearable devices continuously monitor vitals and detect any irregularities in the real-time readings. This real-time monitoring alerts healthcare providers of any necessary action to be taken for potential health risks before they get critical, allowing timely interventions [26]. Smartwatches, bands, or tracking devices such as Radio-Frequency Identification (RFID) bands and other wearable devices have IoT-enabled technologies that track patient vitals such as heart rate, oxygen stats, blood pressure, sleep patterns, patient movements, and activity levels, thus providing insights about the user's health. These devices could also be customized to give alerts about appointments and medication reminders [27].

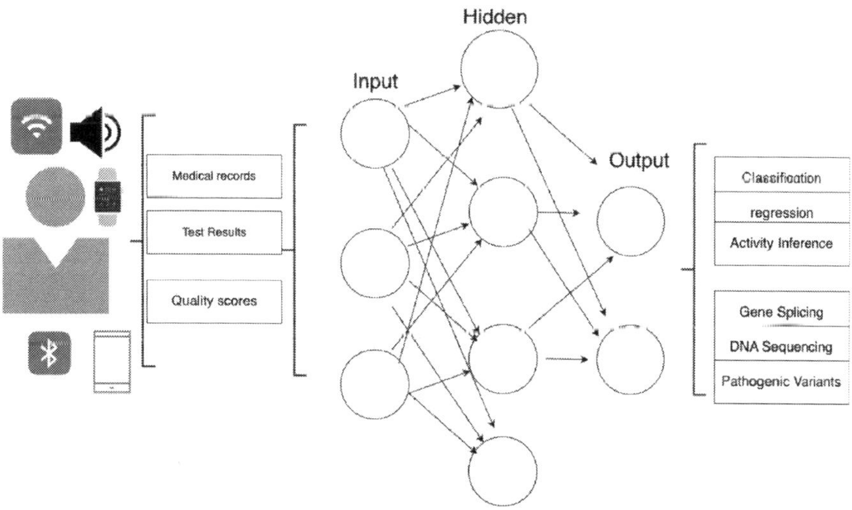


*Figure 10.4* Deep feature extraction and model building for healthcare [26].

### 10.4.3 Industrial IoT

- *Predictive Maintenance and Quality Control*
  In industrial IoT, edge-based DL systems could predict the maintenance timeline of certain machinery by analyzing their current condition and predicting when maintenance is needed. This reduces downtime and keeps the systems running without any breakdowns. This also helps the equipment run smoothly. The quality of the equipment used in manufacturing can be regulated by DL models on the edge. They can inspect the machinery and give feedback. Product inspection by these models could be done in real time during the manufacturing process. This identifies any defects early during the process itself and maintains high quality standards. Automated quality control speeds up production and eliminates human error [28].

### 10.4.4 Smart cities

- *Traffic Management and Surveillance Systems*
  IoT devices are extensively used in logistics, such as recording driving routes in Global Positioning Systems (GPS) applications. In smart cities, edge-based DL can be used to manage traffic flow [29]. For example, in traffic management for smart cities, edge devices can process data from traffic cameras and sensors instantly to adjust traffic lights. Speed-monitoring systems can analyze traffic in real time, offer route suggestions, and optimize traffic flow by controlling traffic lights to help reduce congestion. This can help to reduce traffic jams and smoothen traffic flow in the city. Any anomaly or suspicious activity could be detected using edge devices in surveillance systems by analyzing video feeds in real time. This promotes public safety by allowing quick responses to potential threats by the authorities. These systems provide security and reduce the need for constant human monitoring, making the surveillance systems more efficient [30].

### 10.4.5 Agriculture

- *Precision Farming and Livestock Monitoring*
  Precision farming generally yields more quality products due to its focus on real-time monitoring of crops. Edge-based DL enables precision farming. Sensors in the field create a system that monitors crop health, soil conditions, and moisture levels to activate actuators for automated irrigation or regulate fertilizers to improve crop health more efficiently. Real-time insights into crop conditions help manage resources, leading to sustainable farming practices. Intelligent edge-enabled wearable sensors like health monitors could help track livestock health, fitness, and behavior of animals. This helps manage and maintain their health, allowing farmers to take any necessary action when needed, thus improving productivity [31, 32, 33].

## 10.5 Challenges in implementing DL on edge devices

Implementation of complex DL models on edge devices has its own challenges, these are addressed to harness the full potential of intelligent IoT systems. The following are the main issues that may arise when implementing DL models on the edge [35], as shown in Figure 10.5.

### *10.5.1 Resource constraints*

- *Computational Power and Memory Limitations*
  Edge devices are generally equipped with limited resources like power output and computational abilities. For example, devices like microcontrollers, smartphones, and sensors don't have the same processing ability as big cloud servers. Running complex DL models on these devices can be challenging. Major issues are raised due to limited computational abilities, which results in slower performance and increased energy consumption [36]. Memory is a crucial contributor to a device's performance. These edge devices have limited memory or Random Access Memory (RAM), which makes it difficult to run complex DL models without compromising their performance and accuracy, as these models require substantial memory to execute tasks and run efficiently. It is challenging to store and process these models without compromising performance [32].
- *Model Optimization Techniques*
- *Quantization and Pruning*
  Optimizing the model and reducing its size by converting the weights and activation functions from the level of high precision (like 32-bit) to lower precision (like 16-bit) improves the speed of the model in executing

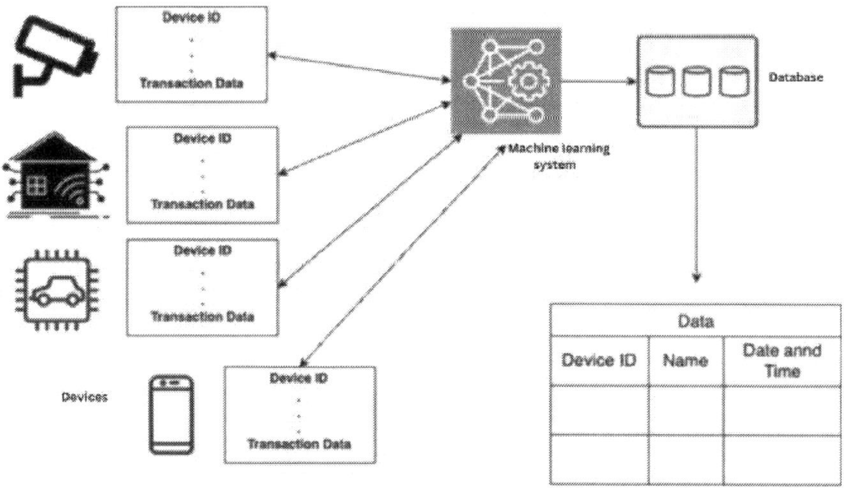


*Figure 10.5* Edge computing Data abstraction [34].

tasks faster without compromising accuracy and performance percentage. Keeping the balance of quantization is the key. Pruning is a technique that removes less important neurons or connections in the neural network, optimizing for faster execution. This speeds up model computation by reducing its size. However, this needs to be done carefully to keep the functionality of the model intact and avoid any degradation in the model's performance [37, 38].

- *Knowledge Distillation*
  When knowledge from a pretrained model is used to train an unlabeled domain, enabling it to replicate the traits of the larger model, this technique is called knowledge distillation. This technique can deploy efficient models on edge devices while retaining much of the original model's accuracy [38].

### 10.5.2 Data management

- *Data Collection and Preprocessing*
  Efficient data management strategies need to be implemented to ensure that data are clean, labeled, and ready for processing. It can be challenging for edge devices to collect and preprocess data as they have limited storage computational capacity [39].
- *Data Privacy and Security*
  Since edge devices are often used for personal and sensitive environments such as homes and healthcare, handling sensitive data on edge devices raises privacy and security concerns. Strong encryption and secure data handling practices are essential to implement [40].

### 10.5.3 Deployment and maintenance

- *Continuous Learning, Updates, and Device Management*
  Keeping edge devices updated can be a little challenging as they need to be individually updated. Unlike cloud servers, which are centrally regulated, newer model versions are pushed for many devices in the field, making it complex and time-consuming. Continued learning also needs monitoring where device models keep improving with new data as presented [41]. Device management can be challenging in a wide network with multiple devices. Each connected device is different, with different operational conditions, capabilities, or firmware versions. Ensuring all the devices run the latest models, function correctly, and securely transmit data requires a robust device management system [33].

## 10.6 Case studies and experimental results

To acknowledge the practical implementation and real-world challenges of DL-based intelligent edge computing systems, we discuss some case studies to analyze and compare edge-based approaches to traditional cloud computing methods. We'll also highlight performance metrics like energy consumption, accuracy, precision, latency, and scalability [17–19], as shown in Figure 10.6 and 10.7.

### 10.6.1 Detailed case studies of edge-based DL applications

#### 10.6.1.1 Smart home security

Cameras in homes could be equipped with DL-based models that detect any intrusions. In smart homes, security systems could process the captured video locally and alert the homeowners instantly without having to receive processed data from the cloud, causing delays. By using this system, there is a significant decrease in response time, hence quick precautions can be taken during a breach [24, 25].

#### 10.6.1.2 Healthcare monitoring

DL models can be used in health monitoring edge devices such as wearable devices where continuous monitoring of patient's vitals is required, especially in case of chronic illnesses where vital signs like heart rate, oxygen stats,

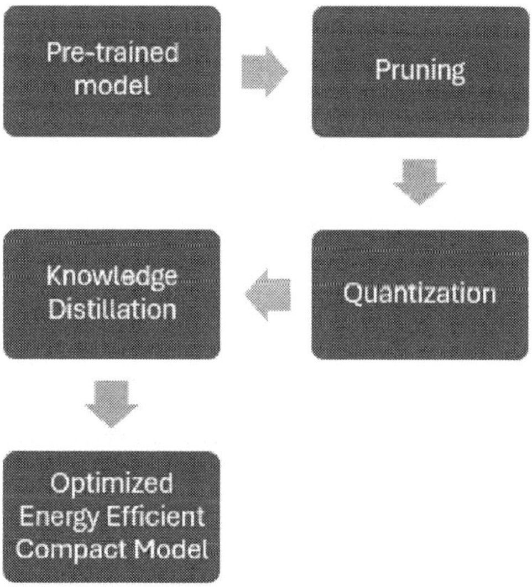


*Figure 10.6* Proposed steps to optimize and deploy a compact model on edge devices.

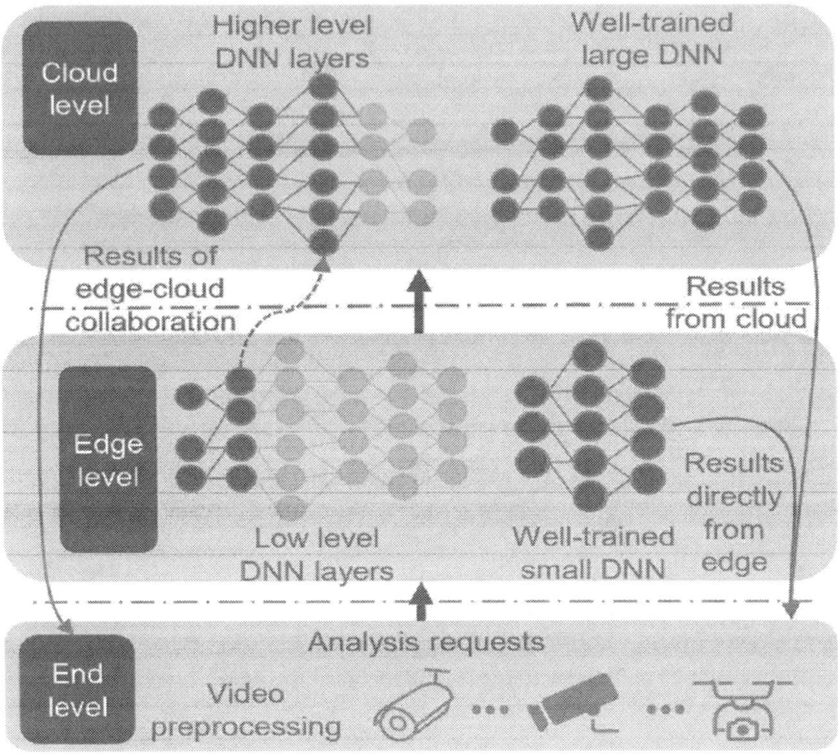


*Figure 10.7* Integration of cloud, edge, and end layer for real-time video analytics using DL [43].

and blood pressure are crucial to indicate the condition. This provides early detection of any potential health issues. This real-time continuous analysis helps in quick response time for medical teams to act, improving patient outcomes and reducing hospital visits. Hospital visits allow timely medical interventions, improving patient outcomes [26].

#### *10.6.1.3 Industrial IoT for predictive maintenance*

IIoT requires machinery sensors that activate the actuators according to the data provided by the sensors in real time. DL-based edge intelligence in IIoT systems also predicts the maintenance needs of the machinery, identifying any signs of wear and tear requiring actions or replacement. A case study at a manufacturing plant showed a reduction in unexpected downtime by 40% as maintenance could be scheduled proactively based on data insights [42].

### 10.6.2 *Comparative analysis with traditional cloud-based approaches and edge-based systems*

Comparative analysis is done in this section, where edge-based systems are compared against cloud-based systems on performance metrics like energy consumption, latency, scalability, accuracy, and precision, where values are considered based on widely recognized trends and evidence [17].

#### *10.6.2.1 Latency*

Edge-based systems significantly reduce latency. As mentioned previously, in traffic management for smart cities, edge devices can process data from traffic cameras and sensors instantly to adjust traffic lights. Compared to cloud-based systems, which may take several seconds to process and respond due to data transmission delays, edge systems respond in milliseconds [17].

#### *10.6.2.2 Energy consumption*

Edge devices are more energy-efficient compared to constantly sending data to the cloud. In precision farming, sensors in the field that use edge computing for data processing consume less power as they avoid continuous data transmission [19, 20].

#### *10.6.2.3 Accuracy, precision, and scalability*

Edge-based models provide and maintain high precision and accuracy. These models are optimized for efficiency. In manufacturing, edge devices, such as cameras, use DL models to detect defects with accuracy comparable to cloud-based devices for quality control applications. Due to local processing, the difference is with the speed at which the detection happens [19, 20]. Edge-based systems are inherently scalable. Adding new devices doesn't strain a central server since each device processes data independently.

In smart cities, deploying additional sensors for monitoring air quality, noise levels, or other parameters is straightforward and does not require extensive infrastructure changes [21].

### 10.6.3 *Performance metrics and evaluation*

#### *10.6.3.1 Latency*

Edge-based systems excel in reducing latency. For instance, in a case study involving autonomous drones for agricultural monitoring, edge processing allowed drones to analyze crop health in real time and make immediate adjustments to their flight paths. Quick responsiveness of such systems is crucial for applications where immediate action is required [17].

#### *10.6.3.2 Energy consumption*

Experiments with smart meters in households for energy management, which use edge-based systems, resulted in a 30% reduction in energy consumption as compared to cloud-based solutions. Since data are processed locally, there is no need to transmit data over the network, making the process less energy intensive [19].

#### *10.6.3.3 Accuracy, precision, and scalability*

A study that compared defect detection in quality control in manufacturing, which uses edge-based systems and cloud-based systems, showed that both approaches yield similar levels of accuracy. However, faster results are obtained by the model using edge-based systems. This highlights the benefits of immediate data processing without compromising accuracy [21]. Edge-based systems proved to be more scalable in a smart city project where hundreds of sensors were deployed to monitor various environmental parameters. Each sensor operated independently, processing data on the edge, which made the system easy to expand without overloading a central server [19, 20].

## 10.7 Conclusion

In this chapter, we explored the advantages and significance of integrating DL with edge computing techniques for the advancement of IoT technologies. There are many limitations of cloud-based systems, which are also addressed here. DL uses learning algorithms to make an IoT system more intelligent, efficient, and secure for data processing. We discussed the architecture, hardware, and software tools, as well as the advantages and practical applications of such models. By processing data locally, edge devices provide real-time insights and quick decision-making capabilities, which are crucial for applications in smart homes, healthcare, industrial IoT, smart cities, and agriculture. We delved into various DL models, such as CNNs, GNNs, and RNNs, and discussed how they can be optimized for edge devices using techniques like quantization, pruning, and knowledge distillation. Specialized hardware like GPUs, TPUs, and dedicated edge AI chips, along with software frameworks like TensorFlow Lite and PyTorch Mobile, enable efficient deployment of these models on the edge. The practical benefits of these systems are highlighted by the experimental results and the case studies discussed. The use cases discussed in the chapter highlight the evolving technology and contribution of edge-based DL in innovating across various industries, from smart cities to agriculture. This chapter provided a comprehensive overview of how ML techniques in the IoT domain revolutionize the industry. From this chapter, we have seen that edge intelligence plays a crucial role in boosting the capabilities of IoT systems and, when combined with smart learning like DL, significantly enhances the effectiveness of IoT systems.

**Acknowledgement:** This work was supported by the Research Council of Finland with CHIST-ERA, Grant agreement no—359790, Di4SPDS-Distributed Intelligence for Enhancing Security and Privacy of Decentralized and Distributed Systems.

## References

[1]. Yoosefi, A., & Kargahi, M. (2024). Resource-aware in-edge distributed real-time deep learning. *Internet of Things*, *27*, 101263.

[2]. Ju, T., Li, L., Liu, S., & Zhang, Y. (2024). A multi-UAV assisted task offloading and path optimization for mobile edge computing via multi-agent deep reinforcement learning. *Journal of Network and Computer Applications*, *229*, 103919.

[3]. Elhanashi, A., Dini, P., Saponara, S., & Zheng, Q. (2023). Integration of deep learning into the iot: A survey of techniques and challenges for real-world applications. *Electronics*, *12*(24), 4925.

[4]. Li, N., Zhai, L., Ma, Z., Zhu, X., & Li, Y. (2024). Lyapunov-guided deep reinforcement learning for service caching and task offloading in mobile edge computing. *Computer Networks*, *250*, 1–12.

[5]. Shi, W., Cao, J., Zhang, Q., Li, Y., & Xu, L. (2016). Edge computing: Vision and challenges. *IEEE Internet of Things Journal*, *3*(5), 637–646.

[6]. Khan, L. U., Yaqoob, I., Tran, N. H., Kazmi, S. A., Dang, T. N., & Hong, C. S. (2020). Edge-computing-enabled smart cities: A comprehensive survey. *IEEE Internet of Things Journal*, *7*(10), 10200–10232.

[7]. Dave, R., Seliya, N., & Siddiqui, N. (2021). The benefits of edge computing in healthcare, smart cities, and IoT. arXiv preprint arXiv:2112.01250.

[8]. Kumar, P., Gupta, G. P., & Tripathi, R. (2023). A review on intrusion detection systems and cyber threat intelligence for secure IoT-enabled networks: Challenges and directions. In *Big Data Analytics in Fog-Enabled IoT Networks* (pp. 51–76). Taylor and Francis.

[9]. Voghoei, S., Tonekaboni, N. H., Wallace, J. G., & Arabnia, H. R. (2018, December). Deep learning at the edge. In *2018 International Conference on Computational Science and Computational Intelligence (CSCI)* (pp. 895–901). IEEE.

[10]. Gill, S. S., Golec, M., Hu, J., Xu, M., Du, J., Wu, H., . . . Uhlig, S. (2024). Edge AI: A taxonomy, systematic review and future directions. arXiv preprint arXiv:2407.04053.

[11]. Aljuhani, A., Kumar, P., Alanazi, R., Albalawi, T., Taouali, O., Islam, A. N., . . . Alazab, M. (2023). A deep learning integrated blockchain framework for securing industrial IoT. *IEEE Internet of Things Journal*, *5*.

[12]. Krizhevsky, A., Sutskever, I., & Hinton, G. E. (2012). ImageNet classification with deep convolutional neural networks. In *Advances in Neural Information Processing Systems* (p. 25). ACM.

[13]. Kipf, T. N., & Welling, M. (2016). Semi-supervised classification with graph convolutional networks. arXiv preprint arXiv:1609.02907.

[14]. Sutskever, I. (2014). Sequence to sequence learning with neural networks. arXiv preprint arXiv:1409.3215.

[15]. Zhuang, F., Qi, Z., Duan, K., Xi, D., Zhu, Y., Zhu, H., . . . He, Q. (2020). A comprehensive survey on transfer learning. *Proceedings of the IEEE*, *109*(1), 43–76.

[16]. Dutta, L., & Bharali, S. (2021). Tinyml meets IoT: A comprehensive survey. *Internet of Things*, *16*, 100461.

[17]. Su, W., Li, L., Liu, F., He, M., & Liang, X. (2022). AI on the edge: A comprehensive review. *Artificial Intelligence Review*, *55*(8), 6125–6183.
[18]. Ali-Eldin, A., Wang, B., & Shenoy, P. (2021, November). The hidden cost of the edge: A performance comparison of edge and cloud latencies. In *Proceedings of the International Conference for High Performance Computing, Networking, Storage and Analysis* (pp. 1–12). ACM.
[19]. Li, Q., Li, H., & Meng, L. (2024). A generic deep learning architecture optimization method for edge device based on start-up latency reduction. *Journal of Real-Time Image Processing*, *21*(4), 116.
[20]. Kong, X., Wu, Y., Wang, H., & Xia, F. (2022). Edge computing for internet of everything: A survey. *IEEE Internet of Things Journal*, *9*(23), 23472–23485.
[21]. Shi, W., Cao, J., Zhang, Q., Li, Y., & Xu, L. (2016). Edge computing: Vision and challenges. *IEEE Internet of Things Journal*, *3*(5), 637–646.
[22]. Bhat, S. A., Sofi, I. B., & Chi, C. Y. (2020). Edge computing and its convergence with blockchain in 5G and beyond: Security, challenges, and opportunities. *IEEE Access*, *8*, 205340–205373.
[23]. Alsharif, M. H., Kelechi, A. H., Jahid, A., Kannadasan, R., Singla, M. K., Gupta, J., & Geem, Z. W. (2024). A comprehensive survey of energy-efficient computing to enable sustainable massive IoT networks. *Alexandria Engineering Journal*, *91*, 12–29.
[24]. Ben Sada, A., Naouri, A., Khelloufi, A., Dhelim, S., & Ning, H. (2023). A context-aware edge computing framework for smart internet of things. *Future Internet*, *15*(5), 154.
[25]. Sulieman, N. A., Ricciardi Celsi, L., Li, W., Zomaya, A., & Villari, M. (2022). Edge-oriented computing: A survey on research and use cases. *Energies*, *15*(2), 452.
[26]. Thakur, P., Goel, S., & Puthooran, E. (2024). Edge AI enabled IoT framework for secure smart home infrastructure. *Procedia Computer Science*, *235*, 3369–3378.
[27]. Moallem, M., Hassanpour, H., & Pouyan, A. A. (2019). Anomaly detection in smart homes using deep learning. *Iranica Journal of Energy & Environment*, *10*(2), 126–135.
[28]. Ramanathan, S., Pineda-Briseno, A., Mohd, T. K., & Ramasundaram, M. (2024). Edge computing in healthcare: Concepts, tools, techniques, and use cases. In *Handbook of AI-Based Models in Healthcare and Medicine* (pp. 1–18). CRC Press.
[29]. Ikonen, J., Nelimarkka, N., Nardelli, P. H., Mattila, N., & Melgarejo, D. C. (2022). Experimental evaluation of end-to-end delay in a sigfox network. *IEEE Networking Letters*, *4*(4), 194–198.
[30]. Wan, J., AAH Al-awlaqi, M., Li, M., O'Grady, M., Gu, X., Wang, J., & Cao, N. (2018). Wearable IoT enabled real-time health monitoring system. *EURASIP Journal on Wireless Communications and Networking*, *2018*(1), 1–10.
[31]. Hafeez, T., Xu, L., & Mcardle, G. (2021). Edge intelligence for data handling and predictive maintenance in IIOT. *IEEE Access*, *9*, 49355–49371.
[32]. Guo, M., Pissinou, N., & Iyengar, S. S. (2019, October). Privacy-preserving deep learning for enabling big edge data analytics in Internet of Things. In *2019 Tenth International Green and Sustainable Computing Conference (IGSC)* (pp. 1–6). IEEE.
[33]. Gochhayat, S. P., Lal, C., Sharma, L., Sharma, D. P., Gupta, D., Saucedo, J. A. M., & Kose, U. (2020). Reliable and secure data transfer in IoT networks. *Wireless Networks*, *26*(8), 5689–5702.

[34]. Bhat, S. A., Sofi, I. B., & Chi, C. Y. (2020). Edge computing and its convergence with blockchain in 5G and beyond: Security, challenges, and opportunities. *IEEE Access*, *8*, 205340–205373.

[35]. Liu, G., Shi, H., Kiani, A., Khreishah, A., Lee, J., Ansari, N., . . . Yousef, M. M. (2021). Smart traffic monitoring system using computer vision and edge computing. *IEEE Transactions on Intelligent Transportation Systems*, *23*(8), 12027–12038.

[36]. Alonso, R. S., Sittón-Candanedo, I., García, Ó., Prieto, J., & Rodríguez-González, S. (2020). An intelligent edge-IoT platform for monitoring livestock and crops in a dairy farming scenario. *Ad Hoc Networks*, *98*, 102047.

[37]. Naveen, S., & Kounte, M. R. (2023). Compact optimized deep learning model for edge: A review. *International Journal of Electrical & Computer Engineering*, *13*(6), 6904–6912.

[38]. Li, X., Li, Y., Li, Y., Cao, T., & Liu, Y. (2024, May). FlexNN: Efficient and adaptive DNN inference on memory-constrained edge devices. ACM.

[39]. Marinó, G. C., Petrini, A., Malchiodi, D., & Frasca, M. (2023). Deep neural networks compression: A comparative survey and choice recommendations. *Neurocomputing*, *520*, 152–170.

[40]. Varghese, B., Wang, N., Barbhuiya, S., Kilpatrick, P., & Nikolopoulos, D. S. (2016, November). Challenges and opportunities in edge computing. In *2016 IEEE International Conference on Smart Cloud (SmartCloud)* (pp. 20–26). IEEE.

[41]. Gaikwad, S. V., Vibhute, A. D., Kale, K. V., & Mehrotra, S. C. (2021). An innovative IoT based system for precision farming. *Computers and Electronics in Agriculture*, *187*, 106291.

[42]. Liang, F., Yu, W., Liu, X., Griffith, D., & Golmie, N. (2020). Toward edge-based deep learning in industrial Internet of Things. *IEEE Internet of Things Journal*, *7*(5), 4329–4341.

[43]. Wang, X., Han, Y., Leung, V. C., Niyato, D., Yan, X., & Chen, X. (2020). Convergence of edge computing and deep learning: A comprehensive survey. *IEEE Communications Surveys & Tutorials*, *22*(2), 869–904.

Chapter 11

# Hybrid deep learning–based edge intelligence and analytics

*K. Ishwarya and Joseph James S.*

## 11.1 Introduction

There has been a recent shift towards a data-driven society, with end devices producing massive amounts of data that can be used for development purposes, such as training and inference [1–3]. The fields of computer vision, large data analysis, and natural language processing (NLP) have all witnessed encouraging advancements in deep learning (DL) as of late. As an example, DL models like Gropher, BERT, Megatron-LM, and GPT-3 are achieving comprehension of textual input in NLP tasks comparable to that of humans [4]. On a number of occasions, DL models have outperformed humans. This includes game-like real-time strategy situations [5] and object categorization tasks [6, 7]. Most use cases for DL training and deployment involve a centralized cloud-based architecture. For many mission-critical use cases, however, the transport, processing, and collection of massive amounts of data to the central cloud become a bottleneck [8, 9]. Edge computing, in this sense, connects on-premises systems to cloud services, both public and private, with remarkable speed. The network's periphery, which typically has limited hardware and memory capabilities, might provide crucial support for DL at the periphery.

An emerging paradigm called "Edge Intelligence and Analytics" integrates the strengths of "edge computing," "AI," and "advanced analytics" to process, analyze, and extract insights from data at the very edge of the network, in close proximity to its point of generation. Optimizing network bandwidth, reducing latency, improving data privacy, and enabling real-time decision-making are all benefits of this method. One new subfield of artificial intelligence (AI), "edge deep learning" targets low-powered devices like smartphones and Internet of Things (IoT) nodes with DL models preloaded onto them. Problems with latency, bandwidth, privacy, and energy efficiency are some of the issues brought up by AI in the cloud, and this paradigm change seeks to resolve them.

One of the best ways to get data out of IoT devices and use them for decision-making is to feed them into a machine learning (ML) or DL system.

DOI: 10.1201/9781003512066-11

The implementation of ML algorithms is impeded by the processing capability of these devices. The standard practice has always been to transfer all data to a central cloud server for processing. The data must be sent and stored eternally on the cloud, which can lead to issues with user data privacy, bandwidth saturation, and high latency. Furthermore, it is becoming more and more difficult to dump all data to the cloud due to the ever-increasing data acquired by each device and the ever-expanding IoT sector.

Lots of different ideas have been put out to fix these problems. Keeping processing near to the point of origin and just sending necessary data to distant servers is one efficient approach. This method has many benefits; it's called edge computing or edge intelligence.

The concept of edge intelligence (EI) suggests moving data processing closer to the point of data collection, rather than to a remote location, to the periphery of the IoT network. This includes gateways, embedded devices, and edge servers. Spreading computation across the layers of the network, which involves positioning computer equipment close to the data-generating devices, accomplishes this. To illustrate, the IoT gateway might perform preliminary data processing and then send the interim results to the cloud, where further processing can be carried out. Data dimensionality could be decreased by incorporating ML capabilities into edge computing. Reduced bandwidth needs and workload in cloud systems would result from preprocessing occurring at the edge node to extract crucial data before transmission to the cloud for advanced analysis. The solution not only protects user privacy but also reduces bandwidth consumption by handling all data locally. In response to the merging of DL and edge computing, a new model of intelligence known as EI has emerged [10, 11]. By bringing DL closer to the data source, EI hopes to make its implementation easier. Since EI makes the most of the resources accessible at end devices, edge servers, and cloud servers for DL training and inference, it maximizes efficiency.

## 11.2 The internet of things

A network of interconnected computing devices, sensors, and software that can transfer data between computers and other devices is known as the Internet of Things (IoT). The IoT is commonly imagined as a network of interconnected wireless devices that can collect and transmit data about our physical surroundings. The IoT was once known as "machine to machine" (M2M). Industrial machinery and processes embraced it because it automated control operations involving sensors connected to different computing systems. Through the introduction of models that can generate predictions or choices independently of human involvement, ML provides the opportunity to further develop automation. Traditionally, when integrating IoT and ML, a central server would receive raw sensor data and then do the inference calculations needed to produce a forecast. This is because many ML algorithms are complex.

With simple models and little amounts of raw data, this setup might work. But there are a few problems that could crop up, as follows:

- Network bandwidth can be used up when sending huge amounts of sensor data, like images.
- Data transmission also necessitates energy consumption.
- To provide ML calculations in near real time, the sensors must be in continual communication with the server.

At the "edge," devices that are geographically close to the user or sensor can store data and do certain computations, reducing the burden of transmitting massive volumes of raw data over networks. The term "edge computing" describes any processing that does not take place in a central cloud. Contrast this with cloud computing, which makes remote data and services accessible to users whenever they need them; this is known as edge computing. At the periphery of computing lie both embedded systems (such as the ones that make up the IoT) and mobile devices like smartphones and personal computers. We employ edge machine learning to train all of these devices to be smarter and less dependent on servers in the backend. ML model training using data from IoT devices deployed in cloud is shown in Figure 11.1.

## 11.3 Edge computing

As an expansion of cloud computing, edge computing employs network devices—like smartphones or home gateways—near the end devices—sensors embedded in smart objects—to gather data about the surrounding environment. It is a model for decentralized computing that facilitates computation close to data sources. With the rise of ubiquitous network connectivity and data-driven applications, wireless communication system design and management have undergone a sea change with the arrival of Edge Computing (EC). A novel approach to improving system efficiency and decreasing latency

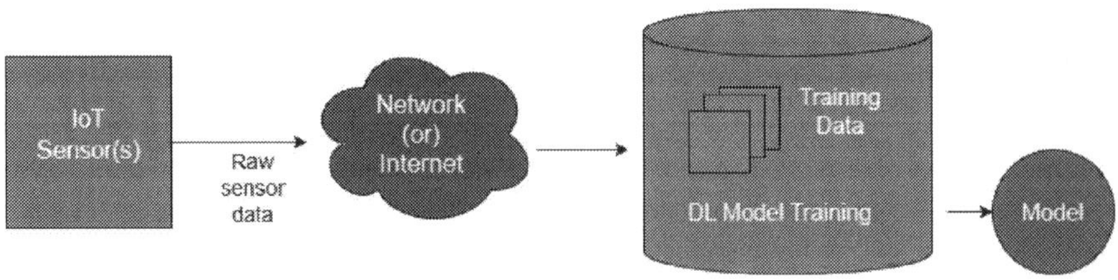


*Figure 11.1* Machine learning in the cloud using data from IoT devices.

has been proposed, which entails moving data processing and computational tasks to edge nodes situated close to end users.

One of the world's most extensive infrastructures for producing data, the IoT is the backbone of smart cities, healthcare, and industry. According to a projection by McKinsey & Company, the IoT would have an annual economic impact of $11.1 trillion in nine different industries by 2025. Innovations in the IoT have resulted in the creation of massive datasets. Data created by the IoT encompasses visuals, sound, and spatial–temporal information. Information is derived via data analytics performed on the sensed data. Typically, analytics are carried out by sending data to a central server or the cloud (Osia et al., 2020; https://www. picturethisai.com/; Unal et al., 2021). According to Shi et al. (2016), traditional methods are being supplanted by edge computing as a result of bandwidth and latency limitations.

## 11.4 Edge and embedded ML

Technological and ML advancements have cleared the way for edge devices to run deep ML models efficiently. Model training, object detection, and NLP are all complex activities that still necessitate powerful computers. This often involves gathering raw data and transmitting it to a server for processing [12]. Multiple advantages, however, can be gained from running ML on low-power devices, as follows:

- The transmission of raw data uses less network bandwidth.
- A smaller amount of communication typically results in lower power consumption, yet some data may require transmission over a network (such as inference findings).
- There is no delay in receiving prediction findings because they have not been transmitted via a network.
- The need for a network connection is not necessary to carry out inference.
- With data stored exclusively for inference purposes (i.e., not including data acquired for model training), user privacy is protected.

Edge Deep Learning encompasses embedded systems, cell phones, and personal computers. Consequently, "edge deep learning" refers to the practice of applying DL algorithms to embedded devices, like microcontrollers and bare-bones single-board computers. ML model training using data from IoT devices deployed in edge devices is shown in Figure 11.2.

For DL models, training typically requires more computing power than inference. Hence, when we need to train new models, we frequently utilize robust server farms. The first step is to build a dataset utilizing data collected from various sources (such as sensors and internet image scraping) and then use that dataset to train our DL model. It is possible to deploy trained models

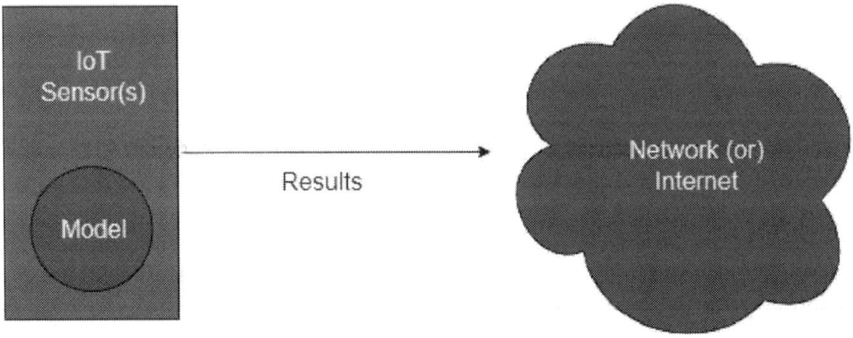


*Figure 11.2* Machine Learning mode deployed in edge devices.

to smart sensors and other edge devices once we have them. These models are essentially mathematical formulas stored in software libraries. Using the model, we may program or modify firmware to collect fresh raw sensor data, draw conclusions, and respond accordingly. Autonomous vehicle operation, robotic arm manipulation, or user notification of a malfunctioning motor are all examples of such tasks. No network connection is required on the edge device since inference is done locally.

## 11.5 Edge deep learning use cases

Edge computing enables a multitude of automation tools and better IoT systems by eliminating the need to constantly connect to a more powerful computer in order to execute machine learning. Some sectors are benefiting from edge ML-enabled innovation, as follows.

### *11.5.1 Agriculture*

#### *11.5.1.1 AI-managed crop irrigation*

In recent times, the use of pesticides, mechanization, and large amounts of water has been the only way to grow enormous quantities of food. On the other hand, these methods wreak havoc on the environment and drain precious resources because they aren't sustainable. Water usage is the target of this research since it is possible to avoid overwatering by identifying when and where water is required. An adaptive approach that uses ML rather than numerous linear regressions or hard thresholds is preferable. More data are also collected over time, which allows for improvements. A decentralized system in which each device operates autonomously is made possible by placing models at the edge instead of in the cloud, where a constant internet connection is necessary.

#### *11.5.1.2 AI-powered agricultural robotics*

A robotic revolution is underway in the agricultural sector. Farmers are getting assistance from agricultural robots when it comes to crop selection and weed eradication. Trained trial robots can do more than just water and spray crops; they can also precisely identify and assess each variety. Agri robots must work with limited power supplies because of their faraway position. The embedded systems installed on them have limited power, which is a problem because navigation and communication are vital systems. Because of their out-of-the-way locations, agricultural robots also require access to contemporary communication networks. The system allows for the transmission of telemetry data and analytical results to the operations center in near real time. This might be a fifth generation (5G) network on the ground or in orbit, depending on where the rollout is taking place. In addition to pinpointing the precise whereabouts of the robot, Global Navigation Satellite System can also be employed to identify specific areas within the crops.

**Precision Farming:** To maximize agricultural yields while minimizing resource use, edge computing processes data in real time from soil sensors, drones, and weather stations to optimize irrigation, fertilization, and pesticide application.

### *11.5.2 Smart buildings*

#### *11.5.2.1 Smart HVAC system*

An intelligent Heating, Ventilation, and Air Conditioning (HVAC) system is more than just a single unit. So, a downstairs workshop that is never used won't have its temperature set to a constant, wasteful level because the HVAC system adapts to the demand in each room. A smart HVAC unit, smart vents, thermostats, and sensors make up the system. In every room, you'll find a single sensor, thermostat, and smart vent. There is two-way communication between the central unit and all of the vents, sensors, and thermostats. The presence of humans or animals in a given space can be detected via sensors. Each room's temperature is detected by the thermostat and transmitted to the central unit as well as the smart device of the customer. Depending on the demand, the vents can be left open or closed. For example, the sensor can detect an individual's presence in Room A. After that, it will communicate with the linked thermostat to confirm that the temperature is what it needs to be. To activate the heating or cooling system, the thermostat communicates with the main unit if the indoor temperature is higher or lower than the set point. In addition, Room A's smart vent will be fully open to allow warm or chilly air to enter the room.

### 11.5.3 Environment conservation

#### 11.5.3.1 Smart grid monitoring system

Some of the most massively linked systems on the planet are the electric power grids that provide energy to the devices that power our contemporary way of life. Predicting and preventing possible outages and failovers is more crucial than ever before because of rising energy demand, changing environmental conditions, and the need to handle varied energy sources including renewable and gas-fired facilities. Even a little breakdown in the infrastructure that distributes high-power electricity can have far-reaching and disastrous effects. Edge Impulse is a cutting-edge smart grid monitoring solution that runs embedded ML algorithms. In order to make the pole-mounted gadgets capable of making inferences efficiently such that they can survive 20 years on a single battery charge, the researchers developed ML models. Power grid operators may now remotely identify various defects and respond swiftly to breakdowns, all while preserving power consumption.

### 11.5.4 Industry

#### 11.5.4.1 AI-automated hard-hat detection

Aside from being the most fragile, the human brain is also the most intricate organ. If we want to keep accidents to a minimum and lessen the frequency of severe brain injuries, we must prioritize risk assessment and prevention in the workplace. As a result, hard hats are frequently mandated by legislation in industrial workplaces; they serve to ward off catastrophes. An object detection model trained using DL was able to identify the wearer's helmet. Use Edge Impulse Studio to collect your own data, make it visually appealing, train your ML model, and check your inferences. It is possible to implement the inference model using a Raspberry Pi and an NVIDIA Jetson Nano.

### 11.5.5 Health and fitness

#### 11.5.5.1 Portable medical devices—early blindness and diabetic retinopathy detection

Globally, the number of cases of blindness caused by retinopathy may be drastically reduced if there was a portable medical gadget that could reliably detect different stages of diabetic retinopathy, even without an internet connection. It is now feasible to create algorithms that can detect or diagnose using just medical devices that are powered by batteries, all thanks to embedded ML. The disease known as diabetic retinopathy disrupts blood vessel function in the retina and other tissues at the back of the eye. Diabetics whose blood sugar levels are not well controlled are at risk. Blindness can occur in the most severe chronic cases of diabetic retinopathy. One ML model in Edge Impulse has the potential to perform inference on edge devices themselves. Between retinal camera pictures of eyes, the algorithm should ideally be able to determine the degree of diabetic retinopathy.

#### *11.5.5.2 Remote patient monitoring*

By processing data in real time from wearable devices or sensors, edge devices can detect problematic situations including irregular heartbeats, blood sugar levels, or respiratory difficulties right away. With edge computing, data may be processed locally, reducing the vital delay in alerting healthcare providers—an essential feature in actual crises.

### *11.5.6 Manufacturing*

- **Predictive Maintenance:** Computing at the edge paves the way for predictive maintenance by allowing the real-time monitoring of machinery and equipment. Edge sensor data analysis allows manufacturers to foresee equipment breakdowns, cutting down on maintenance expenses and downtime.
- **Quality Control:** Devices near the production line's periphery can monitor and analyze data collected by cameras and sensors in real time, helping to eliminate inconsistencies and maximize efficiency.
- **Automation and Robotics:** The ability for automated machinery and robots to make decisions in real time is made possible by edge computing in smart factories. This allows for more flexible manufacturing processes and increases productivity.

## 11.6 Benefits of edge deep learning

- **Reduced Latency:** Time is of the essence in potentially fatal circumstances, and immediate data processing allows for quicker responses.
- **Improved Data Security:** One way to lessen the likelihood of data breaches is to process data locally, rather than sending it over the internet.
- **Scalability:** By decreasing the load on centralized infrastructure, healthcare systems can scale more easily through the deployment of edge devices closer to patients and data sources.
- **Cost Efficiency:** Healthcare organizations can save money on cloud storage and bandwidth by processing data locally.

## 11.7 Challenges of edge deep learning

### *11.7.1 Limited processing power*

- **Constrained Resources:** When contrasted with computing power, memory, and storage capacity found in centralized data centers, edge devices frequently fall short. Consequences for model performance may include a limitation on the complexity of deployable ML models.

- **Model Simplification:** Constrained resources may necessitate simplification of ML models, which in turn may compromise their accuracy or the variety of tasks they are capable of efficiently doing.

### *11.7.2 Scalability issues*

- **Resource Management:** As each edge device must be managed and updated separately, scaling ML models across many devices can be a complicated ordeal. Operational complexity rises, and device-specific model performance may become inconsistent as a result.
- **Deployment Challenges:** It can be time-consuming and laborious to roll out upgrades or new models throughout a distributed network of edge devices, particularly in distant or hard-to-reach areas.

### *11.7.3 Data privacy and security risks*

- **Data Fragmentation:** Although data privacy can be improved with edge computing by storing it locally, data fragmentation due to scattered devices makes it harder to implement consistent security measures and increases the likelihood of data breaches.
- **Device Vulnerability:** Security of ML models and data could be jeopardized since edge devices are more susceptible to physical manipulation, unauthorized access, or cyberattacks than centralized cloud servers.

### *11.7.4 Energy consumption*

- **Power Constraints:** Batteries are a common restricted power source for edge devices, especially those used in mobile or distant applications. It is possible that these devices' battery lives will be shortened due to the energy consumption of running ML models.
- **Efficiency Trade-offs:** Energy conservation measures taken by edge devices might reduce the efficacy of ML models by compromising computing performance in favor of reduced power usage.

### *11.7.5 Integration challenges*

- **Compatibility Issues:** It might be challenging to guarantee consistency and compatibility when deploying ML models on edge devices due to their potentially diverse hardware and software configurations. There may be a lot of tweaking and testing involved in integrating ML models across various edge platforms.

- **Interoperability:** While dealing with various technologies and standards, it can be especially tough to ensure that edge devices communicate and exchange data seamlessly with centralized systems or other edge devices.

### *11.7.6 Latency in model updates*

- **Delayed Model Deployment:** The decentralized structure of the edge computing network can cause delays in updating ML models to reflect new data or insights. This could impact the timeliness of decision-making in situations where these changes are required often.
- **Real-Time Training Limitations:** Improving ML models on edge devices could cause delays because they can't be trained or updated in real time. Instead, they have to be processed in the cloud and then sent to the edge.

ML at the edge typically requires substantially more computing power than polling sensors and transmitting raw data. It is more energy-efficient to do these computations locally rather than sending the raw data to a distant server. The advent of edge ML has opened up new technological frontiers and made possible creative approaches to old challenges.

## 11.8 Conclusion

A paradigm change is underway with the advent of edge intelligence and analytics, which move processing closer to the point of origin of data. With more and more applications requiring real-time insights and decisions, this method is crucial since it allows for faster, more efficient, and safer data processing. Edge intelligence and analytics are going to be huge in many different industries in the future, helping to drive innovation and boost operational efficiency thanks to the advancements in technology. While there are many advantages to using ML in edge computing, particularly for localized and real-time processing, there are also some disadvantages, such as difficulties with resources, scaling, security, and maintenance. When developing and implementing DL solutions for the edge, organizations must thoroughly consider the pros and cons.

The area of edge deep learning is changing at a rapid pace and will have far-reaching effects on AI in the future. Although DL model optimization for edge devices has come a long way, there are still obstacles to overcome in areas such as security, adaptability, and energy efficiency. To fully utilize AI at the edge, further research in these areas is very necessary.

## References

1. Harine Rajashree, R., Sundarakantham, K., Sivasankar, E., Mercy Shalinie, S. 2023. A hybrid deep learning framework for privacy preservation in edge computing. Computers & Security, 129, 103209.

2. Aggarwal, K. M., Mijwil, M., Al-Mistarehi, A. H., Alomari, S., and Gök, M. 2022. Has the future started? The current growth of artificial intelligence, machine learning, and deep learning. Iraqi Journal for Computer Science and Mathematics, 3, no. 1, 115–123.
3. Lv, H., Shi, S., and Gursoy, D. 2022. A look back and a leap forward: A review and synthesis of big data and artificial intelligence literature in hospitality and tourism. Journal of Hospitality Marketing Management, 31, no. 2, 145–175.
4. Dale, R. 2021. GPT-3: What's it good for? Natural Language Engineering, 27, no. 1, 113–118.
5. Wu, B. Hierarchical macro strategy model for MOBA game AI. Proceedings of AAAI Conference on Artificial Intelligence, 33, 1206–1213.
6. Barbu, A., Mayo, D., Alverio, J., Luo, W., Wang, C., Gutfreund, D., Tenenbaum, J., and Katz, B. 2019. ObjectNet: A large-scale bias-controlled dataset for pushing the limits of object recognition models. Proceedings of Advances in Neural Information Processing Systems, 1–11.
7. Patrick, M. K., Adekoya, A. F., Mighty, A. A., and Edward, B. Y. 2022. Capsule networks—A survey. Journal of King Saud University: Computer and Information Sciences, 34, no. 1, 1295–1310.
8. Ray, P. P., Dash, D., and De, D. 2019. Edge computing for Internet of Things: A survey, e-healthcare case study and future direction. Journal of Network and Computer Applications, 140, 1–22.
9. Sulieman, N. A., Celsi, L. R., Li, W., Zomaya, A., and Villari, M. 2022. Edge oriented computing: A survey on research and use cases. Energies, 15, no. 2, 452.
10. Wang, X., Han, Y., Wang, C., Zhao, Q., Chen, X., and Chen, M. 2019. In-edge AI: Intelligentizing mobile edge computing, caching and communication by federated learning. IEEE Network, 33, no. 5, 156–165.
11. Han, S., Pool, J., Tran, J., and Dally, W. 2015. Learning both weights and connections for efficient neural network. Proceedings of Advances in Neural Information Processing Systems, 28, 1135–1143.
12. Alsalemi, A., Himeur, Y., Bensaali, F., and Amira, A. 2022. An innovative edge-based internet of energy solution for promoting energy saving in buildings. Sustainable Cities and Society, 78, Article no. 103571.

Chapter 12

# A multifaceted approach to security in edge IoT

## Threats, solutions, and future perspectives

*Jomina John and John Singh K.*

## 12.1 Introduction

Processing data at or near the network's edge, where it is created, is known as Edge Internet of Things (Edge IoT). This is in contrast to processing it all in a single, centralized data center or cloud. This technology uses edge computing to relocate computational capacity closer to the data sources, such as sensors and IoT devices, in order to deliver faster data processing, lower latency, and better reaction times. The network of physical objects—often referred to as "things"—embedded with sensors, software, and other technologies to communicate and exchange data with other devices and systems over the internet—is known as the Internet of Things (IoT). Edge devices are IoT devices and sensors that produce data. Edge gateways are devices that operate as a middleman between edge devices and the cloud or data center. Edge computing platforms are the hardware and software required to enable computing at the edge. Finally, connectivity comprises networks and dependable and effective communication protocols that link edge devices and gateways to the cloud and to each other. There are many different uses for edge IoT, including smart homes, healthcare, transportation, and agriculture.

Security is a critical concern in Edge IoT because Edge IoT devices are scattered and frequently handle sensitive data. Both during use and transmission, data should be encrypted to guard against unauthorized access and alteration. Sensitive information stored on edge devices should be protected with secure storage methods. Frequent updates are necessary to fix security problems, and a robust authentication and authorization mechanism is necessary to ensure that only authorized people and devices may access the system. Secure protocols should be in place on edge devices for firmware and software updates. To guarantee network level security in edge IoT, secure communication protocols like transport layer security (TLS) and hypertext transfer protocol secure should be utilized to protect data sent between edge devices and other network components. Segment the network to isolate critical systems and minimize the effect of potential security breaches. To ensure physical security, edge devices must be protected from physical manipulation

DOI: 10.1201/9781003512066-12

by tamper-evident and tamper-resistant designs. To prevent undesired physical access, make sure the edge devices are placed in secure regions. To quickly address and mitigate security issues, an incident response plan needs to be kept current. To keep an eye out for any potential security breaches and unusual activities, an intrusion detection system (IDS) is employed. Any device connected to the network must have a unique, verifiable identity in order to prevent undesirable devices from connecting. Regulate access and safely handle user identities and credentials by utilizing identity and access management (IAM) technology. Data integrity can be ensured and the validity of data from edge devices verified using digital signatures and hashing. In order to guarantee compliance with data protection legislation such as General Data Protection Regulation, Health Insurance Portability and Accountability Act, and others relevant to the industry and area, adhere to industry standards and best practices for IoT security, such as those provided by organizations like NIST, ISO, and the IoT Security Foundation.

## 12.2 Major security issues in edge IoT

The unique characteristics of edge IoT—distributed architecture, a wide range of resource-constrained devices, and handling of sensitive data frequently—create a challenging security environment. In order to ensure data and system accessibility, privacy, and accuracy, edge IoT needs to address several security challenges. The following are the main edge IoT security risks that were identified.

- Denial-of-Service (DoS) attacks: Attacks known as denial of service (DoS) aim to take down edge IoT devices by flooding them with excessive traffic, rendering services unusable for authorized users. This may cause interruptions to regular business operations, which could result in lost time and money. The impact of DoS attacks on Edge IoT devices can be lessened by putting policies in place such as rate limitation, traffic filtering, and anomaly detection [1].
- Eavesdropping: Unauthorized communication between edge IoT devices can be intercepted by eavesdropping, which exposes private information to prospective adversaries. Information sent across the network may become less secret as a result of this security risk. In edge IoT contexts, eavesdropping threats can be prevented by utilizing encryption protocols such as secure sockets layer/TLS and establishing secure communication channels [1].
- Jamming: By obstructing wireless communication channels, jamming attacks aim to compromise edge IoT systems by preventing devices from efficiently transmitting or receiving data [2]. The performance and dependability of edge IoT networks may be impacted by this interference, which can cause data loss and communication problems. Jamming attacks on

edge IoT devices can be lessened by using signal modulation, frequency hopping techniques, and signal strength monitoring [1].

- Malware: Malware infects devices and compromises their functionality, posing a serious threat to edge IoT systems [3]. IoT device vulnerabilities can be exploited by malicious software, which can result in system manipulation, illegal access, and data breaches. Protecting the integrity of edge IoT devices and preventing malware infestations can be achieved by implementing frequent software updates, antivirus software, and network segmentation [1].
- Sinkhole Attack: Through the redirection of legitimate traffic to a malicious node under the control of the attacker, sinkhole attacks target edge IoT networks. Data manipulation, interception, and illegal access to private data are possible outcomes of this. In edge IoT contexts, sinkhole attacks can be identified and prevented with the use of intrusion detection systems, secure routing protocols, and network monitoring [1].
- Sybil Attack: A malicious entity uses sybil assaults to assume several false identities in order to trick edge IoT systems and alter network behavior. Unauthorized access, data manipulation, and network congestion may result from this. The impact of Sybil attacks on edge IoT networks can be lessened by putting in place identity verification methods, access control regulations, and anomaly detection algorithms [1].
- Malicious Node Injection: A malicious entity uses sybil assaults to assume several false identities in order to trick edge IoT systems and alter network behavior. Unauthorized access, data manipulation, and network congestion may result from this.
- Unauthorized Access: Several points of unauthorized access indicate vulnerability, highlighting the necessity of strong access control measures to stop unauthorized users from accessing private information and applications [4].
- Side-Channel Attacks: The weaknesses that could lead to side-channel attacks, which could take advantage of information leaking through unintentional channels and endanger the integrity and confidentiality of data processed at the edge [4].
- Data Tampering: The necessity of guaranteeing the integrity of data gathered and processed at the edge in order to prevent unauthorized modifications is highlighted by potential security threats related to data tampering [4].
- Cyberattacks: The potential for cyberattacks to target edge computing systems highlight the necessity of sophisticated security measures to fend against malware injection attacks and other online threats [4].
- Authentication and Authorization Attacks: Strong authentication methods must be implemented to prevent unwanted access to edge devices and services, as the authentication and authorization attacks pose serious risks unique to edge computing platforms [4].

Additionally, there are many obstacles facing the centralized data processing and storage services in edge IoT, which might result in security and privacy lapses. Large-scale potential privacy breaches are possible, particularly if hackers target an edge node and impact several edge devices at once [5]. These breaches fall into one of the following four categories: device abuse, data interception, data impairment, and access violations. Hackers' unauthorized access to and collection of sensor data from linked devices are serious issues. These devices are especially susceptible to hacking because edge devices' size and processing capability impose inherent constraints on security measures. This vulnerability is further exacerbated by the restricted computational resources and insufficient resilience to repeated attacks [6]. The relationships of trust between manufacturers, service providers, and end users are affected by this security lapse. Additional security problems posed by edge computing's hierarchical network design include data dissemination between nodes and vulnerabilities at various communication layers. In edge computing, common threats including denial-of-service, waterhole, eavesdropping, manipulation, and man-in-the-middle target lightweight IoT devices in particular. The distributed network architecture and the participation of several stakeholders in communication flows cause authentication restrictions. Unprecedented levels of cybersecurity challenges are brought about by the hyperconnectedness of IoT devices and the resource-poor characteristics of edge computing. Security precautions are further complicated by the difficulty of recreating the centralized idea of cloud services and its auditability functions in a dispersed edge service environment.

### 12.2.1 Various security issues reported for edge IoT

Numerous studies demonstrate how many security flaws exist in the processing and deployment of edge IoT. The underlying network security in edge IoT has gained widespread attention from the research community. Consequently, a number of recommendations have been made in this approach. The following is a description of a few noteworthy contributions.

Iindombo et al. [7] enumerated several security vulnerabilities present in edge IoT. Edge IoT systems are highly vulnerable to DoS assaults because of their minimal central processing unit (CPU), memory, and storage capacity. There are concerns of data interception during data transmission between IoT devices and the cloud, which could result in data tampering and theft. Difficulties lie in controlling a wide range of dispersed IoT devices with limited hardware capabilities, which makes carrying out intricate calculations challenging. Even though edge computing reduces latency and increases throughput, it has significant drawbacks, including hazards to privacy and security. The smart grid is a prime target for cyberthreats such as phishing, DOS attacks, virus dissemination, and eavesdropping, and more since security is neglected. IoT devices in the smart grid have limited memory and processing power, which leaves them vulnerable to distributed DOS assaults.

Patak et al. [8] covered a number of edge IoT security concerns. Since heterogeneous devices make up most IoT networks, distributed security solutions are needed to handle security issues. IoT networks with lightweight security mechanisms are vulnerable to a variety of attacks, including bad-mouthing, self-promotion, and Sybil attacks, which call for further security precautions. The security and integrity of IoT networks are seriously threatened by the existence of hostile IoT devices and users. Maintaining the integrity of the network depends on ensuring safe and dependable connectivity between blockchain nodes, edge devices, and IoT devices. Ensuring the integrity and confidentiality of data transmitted and stored within the edge IoT network is of utmost importance and presents a significant security risk. Inadequate protocols for assessing trust between networked entities may result in weak points and possible security lapses.

Akmandor et al. [9] examined several edge IoT vulnerabilities. It has been demonstrated that IoT sensors are susceptible to attacks that could leak private data about users or systems. For instance, attacks on surveillance cameras that used login credentials that were made public compromised user privacy. Concerns over data privacy have been raised by situations in which businesses have been accused of gathering, keeping, and marketing consumers' private information without authorization. DDoS assaults have targeted IoT devices; one example is the Mirai malware, which infected a large number of IoT devices and was used to launch attacks in Europe and the United States. The assaults on IoT sensors in commercial and residential settings demonstrate how hackers might use weaknesses in sensor systems to access private data or cause false alarms. Encryption and hashing are two examples of cryptographic algorithms that can be used to enhance the energy consumption of IoT sensors. This makes it difficult to combine security and energy efficiency at the same time.

According to Sairam et al. [10], physical security measures like authentication mechanisms for IoT devices are necessary since edge IoT is vulnerable to physical attacks. Additionally, IoT devices' weaknesses, tiny security measures, and heterogeneity make them targets for cybercriminals. Closer observation of smart devices at the edge of IoT networks is required due to the scalability and cost constraints of traditional centralized IT security solutions. The difficulty of putting security features for the Internet of Things into practice at the network edge because of less powerful devices. It can be difficult to accomplish in a centralized way, thus individual traffic analysis of IoT devices is necessary to understand their typical patterns and aberrant activity. Due to the massive amount of data on the network, IoT devices are also vulnerable to other cyberattacks and general security concerns like DoS attacks.

Several security issues with edge-assisted IoT were brought to light by Shen et al. [11]. These include worries about edge nodes' actual locations, the diverse range of edge services offered by various parties, possible security risks from offloading computational tasks, attackers' fabrication of data, compromised edge nodes that result in malicious attacks like Distributed DoS, privacy violations, and service interruptions. For the sake of data integrity, privacy, and the continued availability of IoT services, among other reasons, edge-assisted IoT system security is essential.

Hafeez et al. [12] noted a number of security issues in edge IoT environments, such as the susceptibility of IoT devices to insignificant attacks, the frequency of insecure IoT devices because of resource limitations during development, the absence of security updates for IoT devices, the danger of device-specific exploits aimed at IoT devices, the possibility of IoT device compromise within small office or home office (SOHO) networks for the purpose of launching large-scale attacks, and the impracticality of traditional security solutions in IoT environments due to resource limitations and device heterogeneity. In order to improve the security posture of edge IoT networks and reduce the dangers associated with criminal activity, these concerns highlight the vital need for novel security solutions like IoT-KEEPER.

Zhang et al. [13] identified a number of security concerns, such as the proliferation of numerous edge nodes, their widespread geographic dispersion, the complexity of the environments in which they operate, and the imperative need to guarantee secure data transmission between edge nodes and cloud servers. The research also highlights how critical it is to gauge node security and deploy a security mechanism that is customized for various forms of data transfer. Virtualizing edge nodes and utilizing a security strategy based on security level measurement are the main components of the suggested (Security-aware and Trustworthy Edge Computing for IoT (STEC-IoT)) solution, which attempts to overcome these difficulties. The strategy seeks to both reduce potential security risks associated with edge devices and improve the security posture of edge computing in IoT scenarios.

Several security limitations were brought out by Ganesh et al. [14], including the vulnerability of IoT devices as a result of resource shortages and poor security design, which could result in cyberattacks and data leaks. Concerns regarding the security of smart devices have been highlighted by the growing interconnection of the physical and digital worlds brought about by IoT applications. Effective IoT device security is further complicated by the dependence on antiquated security measures that have a high processor and memory overhead. The study highlights how crucial it is to create fresh approaches to edge-based IoT security in order to reduce security threats and safeguard private information that is processed and transferred at the edge layer.

## 12.3 Methods to avoid security issues in IoT

Despite the ability of edge IoT to handle sensitive and important data in dispersed and resource-constrained contexts, edge IoT devices are vulnerable to a variety of cyber threats, as previously discussed. In order to reduce these risks, scientists have put forth a number of creative solutions that are suited to the particular needs of edge IoT ecosystems. These solutions cover a wide range of technologies and strategies, from utilizing machine learning for threat mitigation and anomaly detection to integrating blockchain for immutable and distributed security. Efficient and scalable IoT network security can be achieved by using solutions such as virtualizing edge nodes, biometric authentication, and lightweight encryption schemes. This section delves into these cutting edge solutions, providing a comprehensive overview of the current state of research and the practical measures that can be implemented to secure edge IoT environments.

Christophe et al. [15] suggested a technique called SPRINT, which offers a thorough framework for inspection and monitoring, with the goal of addressing security and privacy issues in IoT system services. SPRINT incorporates industry-accepted professional standards to guarantee a dependable and auditable assessment of IoT service environments. It comprises an evaluation methodology (SPRINT-METH) and reference technique (SPRINT-REF) that are based on an extensive examination of security/privacy standards and general IoT standards. Through the combination of security and privacy standards, countermeasures, and measurements, the technique aims to establish a trust environment for IoT business services. In order to improve security, SPRINT further highlights the significance of industry-specific standards as well as ongoing evaluation and observation of IoT services. The SPRINT framework's usefulness and efficacy are demonstrated by its iterative development methodology and real-case validation in industries like railroads.

By building a blockchain layer of security, Iindombo et al.[7] suggested integrating edge computing and blockchain to address vulnerabilities in IoT devices at the edge of the smart grid system. Blockchain technology allows the system to provide distributed trust, immutable records, and secure data sharing among IoT devices. Enhancements such as peer-to-peer messaging, tamper-proof data, and historical activity tracking are made possible by the combination of edge computing and blockchain technology. Enhancing edge IoT device security and resilience in smart grid systems is the ultimate goal of this comprehensive solution.

The Trust-Based ABAC Mechanism, as proposed by Patak et al. [8], integrates access control and trust evaluation mechanisms to improve security by identifying and blocking malicious IoT devices and users from accessing resources. The mechanism is based on blockchain technology. Permissioned Hyperledger blockchain technology is used by the system to record transactions, protect the network, and apply dynamic decentralized trust evaluation for increased dependability and security. Akmandor et al. [9] suggested including ML inference and signal compression on the IoT sensor node as a way to get around the security problems with edge IoT. The quantity of data that needs to be conveyed is greatly decreased by employing compressed-domain inference, which is based on ideas like compressive sensing and compressed signal processing. Comparing these reduced data transmission costs to conventional sense-and-transmit IoT sensors reveals significant energy savings. On the sensor node, encryption and hashing are implemented using energy savings from data compression, guaranteeing the secrecy and integrity of the data. The IoT system can transfer signal processing and decision-making to the edge and improve security and energy efficiency at the same time by utilizing ML inference, hashing, encryption, and compression directly on the sensor node.

Using a lightweight Docker-based solution, Sairam et al. [10] suggested a way to utilize a NETRA architecture to improve IoT security through NFV-based edge traffic monitoring. The primary objective of this approach is to monitor smart devices carefully and effectively fix vulnerabilities by implementing security functions at the network edge. With the use of NFV technology, NETRA seeks to deliver a more effective and affordable security framework for IoT environments while overcoming the drawbacks of conventional centralized IT security procedures. The design shifts security tasks to the edge, uses real device traffic to train machine learning models for accurate attack detection, and minimizes message overhead to detect and mitigate security threats more quickly. With NETRA's approach, security, efficiency, and threat mitigation in IoT networks are improved via a secure platform that combines SDN and NFV. All things considered, NETRA provides a complete solution for the network edge security of IoT devices, guaranteeing enhanced security and monitoring capabilities.

In order to improve malicious detection capabilities, Shen et al. [11] suggested integrating energy monitoring with security schemes as a solution to security challenges in edge-assisted IoT. ML approaches can extract energy attributes, identify consumption patterns, and infer device conditions by tracking the energy consumption of edge nodes and IoT devices. This method seeks to identify anomalous energy usage linked to malicious or compromised edge nodes, offering a fresh viewpoint on improving security in edge-assisted IoT.

Hafeez et al. [12] presented the IoT-KEEPER method, which employs fuzzy C-Means clustering and fuzzy interpolation scheme to distinguish between benign and malicious network activity produced by IoT devices. The method uses an anomaly detection technique for real-time traffic analysis at edge gateways. In order to stop the detected IoT device from attacking other devices or services, IoT-KEEPER automatically imposes network access restrictions on it upon detecting malicious activity. IoT-KEEPER detects malicious network activity with high accuracy and low false positive rates by leveraging unlabeled traffic data for model training and a proprietary feature analysis technique. The system is a promising solution for improving the security of edge IoT networks because of its small resource footprint, capacity to function effectively on constrained gateway resources, and ability to detect and mitigate a variety of network attacks without the need for complex hardware or explicit attack signatures.

In order to improve security, Zhang et al. [13] suggested virtualizing edge nodes and dividing them into separate virtual networks according to their roles. This approach tries to improve security by establishing an organized and controllable framework. Moreover, the implementation of a security mechanism that relies on security level measurement facilitates the evaluation and assignment of suitable security measures to various categories of data transmissions. By ensuring that various security needs are successfully met, this customized security method helps to mitigate potential vulnerabilities in the system. The effectiveness of the suggested solution is proven through simulation experiments, demonstrating how virtualizing edge nodes and putting in place a security strategy that takes into account the various security requirements of data transmissions in these environments can improve edge computing security in the IoT. To successfully handle security challenges in edge IoT, Ganesh et al. [14] designed a cognitive trust management system. By regulating service level beliefs, fostering trust among IoT devices, and guaranteeing Quality of Service (QoS) standards, this system seeks to improve security. The overall security of an IoT edge device can be enhanced by mitigating malicious behaviors and security threats through the implementation of a cognitive trust management architecture. The system uses QoS characteristics to identify potentially dangerous edge nodes in computing networks at the edge, evaluating trustworthiness based on packet delivery ratio and jitter. Enhancing security and trust among edge-layer IoT devices is a promising outcome of the proposed method, which incorporates cognitive trust evaluation models that take QoS characteristics into account.

## 12.4 Future research directions in security of edge IoT

Future research directions in the field of security for edge IoT should address several critical challenges. IoT devices that are gateways connected to the internet do not yet have extensive security features, especially when it comes to network administration and authentication certification. To strengthen the overall security of IoT systems, it is therefore essential to improve the security mechanisms at these gateways and measure real-time security using mechanisms like Fuzzy-based trust calculation [16], parameters-based trust calculation, and so forth. In order to preserve processing efficiency and avoid device failures, it is imperative to solve concerns related to consensus algorithms and system throughput, considering the significant quantity of IoT devices that are connected. Blockchain technology and machine learning present viable answers to these problems. In order to guarantee scalability and security, lightweight encryption algorithms adapted to the resource limitations of IoT devices must be created. Lightweight authentication techniques should be the subject of future study in order to protect sensitive data breaches and enhance network efficiency.

Another interesting approach is biometric security, which verifies identity through physiological or behavioral traits. For user authentication, researchers need to assess the stability and cost-effectiveness of biometric systems. It is crucial to create an IoT ecosystem in order to standardize and encourage compliance with security technologies. IoT security will be strengthened even more by taking authorization and access control into account and speeding up security algorithm computation. Safeguarding IoT data storage is still a top concern, especially when using blockchain and cloud computing technologies. Future studies must address the privacy concerns related to public blockchain transactions as well as the expenses related to cloud computing.

Furthermore, edge computing poses new difficulties because of the low power capacity of edge devices, even though it is effective in reducing spoofing attacks. Attackers can take advantage of these restrictions by carrying out actions that shorten battery life. In order to advance IoT ecosystem security, it will be imperative to address these vulnerabilities and enhance the resilience of edge IoT devices, as shown in Table 12.1.

## 12.5 Conclusion

Edge IoT, which processes data closer to its source, offers a revolutionary chance to improve the responsiveness and efficiency of IoT systems. To fully achieve its potential, this paradigm change does, however, also present important security challenges that need to be resolved. This study has examined a number of edge IoT security issues, emphasizing the risks and weaknesses present in distributed systems, devices with limited resources, and the management of sensitive data. It is clear from a thorough assessment of current security problems and solutions put forth by different researchers that securing edge IoT environments necessitates a multipronged strategy. Enhancing gateway security, using machine learning and blockchain to handle consensus method problems and system throughput, and creating authentication and encryption schemes that are lightweight and appropriate for devices with limited resources are some of the main areas of study. A thorough security plan must also include the incorporation of biometric security, the creation of IoT security guidelines, and the improvement of data storage security via blockchain and cloud technologies.

The necessity for continuous innovation in security measures is highlighted by future research directions, which include the creation of reliable edge computing solutions that take power constraints into consideration, scalable and effective security protocols, and affordable biometric authentication systems. Enhancing the security, privacy, and resilience of edge IoT systems will be necessary to enable their widespread adoption and trust. Therefore, addressing these research problems will be critical. In summary, even though edge IoT has many benefits, its security environment is dynamic and ever-changing. To reduce the dangers and guarantee that edge IoT devices

*Table 12.1* Comparative Analysis of Trust Evaluation Methods

| *Proposed Method* | *Technology Used* | *Scope* | *Level of Security Achieved* |
|---|---|---|---|
| SPRINT | Professional Standards, Security and Privacy Requirements, Countermeasures, Metrics | Comprehensive assessment and monitoring framework for IoT services | High—Continuous assessment and monitoring of IoT services |
| Blockchain and Edge Computing Integration | Blockchain, Edge Computing | Enhancing security of IoT devices at the edge of smart grid systems | High—Distributed trust, tamper-proof data, safe data sharing |
| Trust-Based ABAC Mechanism | Blockchain (Hyperledger), Access Control, Trust Evaluation Mechanisms | Enhancing security by detecting and isolating malicious IoT devices | High—Improved security and reliability |
| Compressed-domain Inference | Signal Compression, Machine Learning Inference, Encryption, Hashing | Enhancing security and energy efficiency in IoT sensor nodes | High—Data confidentiality and integrity with energy efficiency |
| NETRA Architecture | NFV (Network Function Virtualization), Docker-based Edge Traffic Analysis | Enhancing IoT security through NFV-based edge traffic analysis | High—Efficient and cost-effective security framework |
| Energy Monitoring with Security Schemes | Energy Monitoring, Machine Learning | Enhancing security by detecting malicious activities through energy consumption patterns | Medium—Enhanced malicious detection capabilities |
| IoT-KEEPER | Anomaly Detection, Fuzzy C-Means Clustering, Fuzzy Interpolation | Real-time traffic analysis and anomaly detection at edge gateways | High—High accuracy and low false positive rates in detecting malicious activities |
| STEC-IoT | Virtualization, Security Level Measurement | Enhancing security of edge computing by virtualizing edge nodes and tailored security approach | High—Structured and manageable security framework |

(*Continued*)

*Table 12.1* (Continued)

| *Proposed Method* | *Technology Used* | *Scope* | *Level of Security Achieved* |
|---|---|---|---|
| Cognitive Trust Management System | Cognitive Trust Evaluation, QoS Parameters, Packet Delivery Ratio, Jitter | Enhancing security by establishing trust among IoT devices and managing QoS standards | High Improved security and trust among IoT devices at the edge layer |

can function safely and effectively in increasingly linked contexts, research and development must continue. By tackling the recognized security issues and utilizing cutting-edge technology, we can establish a safe framework for edge IoT in the future.

## References

1. Pawlicki, M., A. Pawlicka, R. Kozik, and M. Choras´. "The Survey and Meta-Analysis of the Attacks, Transgressions, Countermeasures, and Security Aspects Common to the Cloud, Edge, and IoT." *Neurocomputing* 551 (2023). https://doi.org/10.1016/j.neucom.2023.126533.
2. Ramasamy, J., and J. S. Kumaresan. "Image Encryption and Cluster-Based Framework for Secured Image Transmission in Wireless Sensor Networks." *Wireless Personal Communications* 112 (2020): 1355–1368.
3. John, J., and J. Norman. "Analysis of Worm Attack Detection Methods in Cloud: Finding a Better Solution for Malware in Cloud." *International Journal of Applied Engineering Research* 10 (2015).
4. Ghadiri, R., and M. Elhajj. "Security and Performance Analysis of Edge Computing in IoT." In *Proceedings of COMNETSAT 2023*, 542–548. Institute of Electrical and Electronics Engineers Inc., 2023. https://doi.org/10.1109/COMNETSAT59769.2023.10420709.
5. Shynu, P., and K. J. Singh. "Privacy Preserving Secret Key Extraction Protocol for Multi-Authority Attribute-Based Encryption Techniques in Cloud Computing." *International Journal of Embedded Systems* 10 (2018): 287–300.
6. John, J., and J. Norman. "Major Vulnerabilities and Their Prevention Methods in Cloud Computing." In *Advances in Big Data and Cloud Computing: Proceedings of ICBDCC18*, 11–26. Springer, 2019.
7. Iindombo, S., and M. Bere-Chitauro. "An Analysis of the Implementation of Blockchain Technology in Smart Grid Edge IoT Devices." In *Proceedings of ETNCC 2023*, 1–5, 2023. https://doi.org/10.1109/ETNCC59188.2023.10284952.
8. Pathak, A., I. Al-Anbagi, and H. J. Hamilton. "TABI: Trust-Based ABAC Mechanism for Edge-IoT Using Blockchain Technology." *IEEE Access*, 2023.
9. Akmandor, A. O., H. Yin, and N. K. Jha. "Smart, Secure, Yet Energy-Efficient, Internet-of-Things Sensors." *IEEE Transactions on Multi-Scale Computing Systems* 4 (2018): 914–930. https://doi.org/10.1109/TMSCS.2018.2864297.
10. Sairam, R., S. S. Bhunia, V. Thangavelu, and M. Gurusamy. "Netra: Enhancing IoT Security Using NFV-Based Edge Traffic Analysis." *IEEE Sensors Journal* 19 (2019): 4660–4671. https://doi.org/10.1109/JSEN.2019.2900097.

11. Shen, S., K. Zhang, Y. Zhou, and S. Ci. "Security in Edge-Assisted Internet of Things: Challenges and Solutions." 2020. https://doi.org/10.1007/s11432-019-2906-y.
12. Hafeez, I., M. Antikainen, A. Y. Ding, and S. Tarkoma. "IoT-Keeper: Detecting Malicious IoT Network Activity Using Online Traffic Analysis at the Edge." *IEEE Transactions on Network and Service Management* 17 (2020): 45–59. https://doi.org/10.1109/TNSM.2020.2966951.
13. Zhang, P., C. Jiang, X. Pang, and Y. Qian. "STEC-IoT: A Security Tactic by Virtualizing Edge Computing on IoT." *IEEE Internet of Things Journal* 8 (2021): 2459–2467. https://doi.org/10.1109/JIOT.2020.3017742.
14. Ganesh, D., K. Suresh, M. S. Kumar, K. Balaji, and S. Burada. "Improving Security in Edge Computing by Using Cognitive Trust Management Model." In *Proceedings of ICECAA 2022*, 524–531. Institute of Electrical and Electronics Engineers Inc., 2022. https://doi.org/10.1109/ICECAA55415.2022.9936568.
15. Feltus, C., T. Grandjean, J. Aubert, and D. Khadraoui. "Towards a Standard-Based Security and Privacy of IoT System's Services." In *Proceedings of CSCI 2018*, 1036–1039. Institute of Electrical and Electronics Engineers Inc., 2018. https://doi.org/10.1109/CSCI46756.2018.00201.
16. John, Jomina, and John K. Singh. "Trust Value Evaluation of Cloud Service Providers Using Fuzzy Inference Based Analytical Process." *Scientific Reports* 14, no. 18028 (2024). https://doi.org/10.1038/s41598-024-69134-8.

Chapter 13

# Case study

## Developing edge intelligence in electric vehicles

*Madhukrishna Priyadarsini and Padmalochan Bera*

### 13.1 Introduction

#### *13.1.1 Context and importance of edge intelligence in electric vehicles*

**Overview of electric vehicles and their rapid adoption:** Due to increase of carbon emissions and cost of natural oils, the next generation is going to face a tremendous transportation challenge. As a result of the same, Electric Vehicles (EVs) are gaining popularity because they are free from greenhouse gas effects [1]. In addition, EV's "tank-to-wheels" concept is three times more powerful than internal combustion engines (ICVs). At the same time EVs are reducing noise and vibrations in the vehicles. As EVs are tackling climate change challenges and contributing toward sustainable energy development, majority of the globe is promoting it. The evolution of EVs is presented in Figure 13.1.

Due to the popularity of EVs, the fastest growing country India is aiming for EV adoption to reach 40% for buses, 30% for private cars, 70% for commercial vehicles, and 80% for two-wheelers by 2030 including pure EVs and plug-in hybrid cars (per NITI Aayog report). In addition, to encourage the growth of charging stations, the Indian government has launched several schemes to incentivize the development of alternative fuel infrastructure, such as subsidies and grants. The current market trends of EVs in India is shown in Figure 13.2.

Despite all the benefits the EVs are providing, they are only capturing 14% of the whole automobile market. Mostly the users are not switching to EVs due to lack of charging infrastructures and the issues present in the undeveloped battery technologies [2]. High-quality solutions are required to solve the issues and to capture the automobile market in order to establish the EV industry.

**Introduction to edge intelligence and its relevance to EVs:** Edge intelligence refers to the processing and analysis of data as close as possible to the data

DOI: 10.1201/9781003512066-13

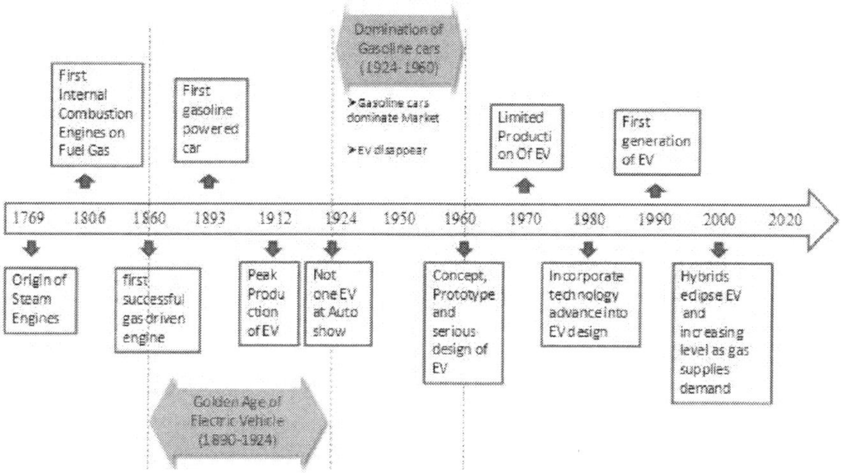


*Figure 13.1* Evolution of EVs.

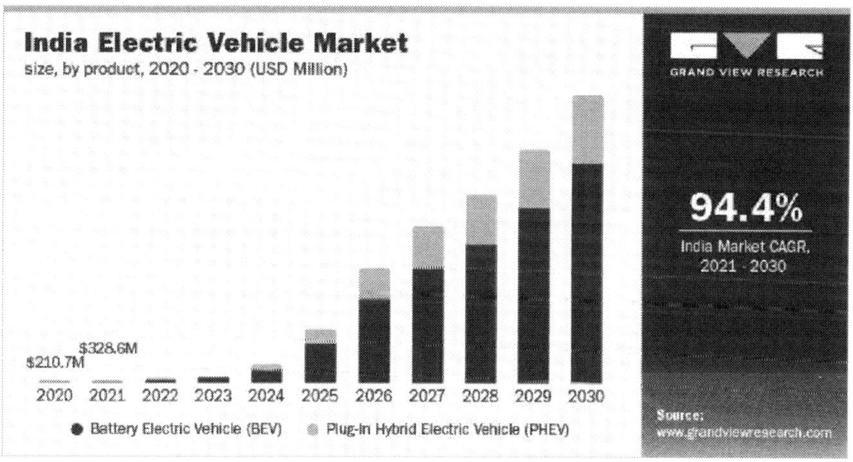


*Figure 13.2* Evolution of EVs in India.

sources rather than depending on the cloud. In terms of EV industry, the vehicles sensors need to process and analyze the data collected by them locally in order to make real-time decisions and actions. *Edge Computing*, *Artificial Intelligence* (AI), *Internet of Things* (IoT) are the three key components of edge intelligence, where edge computing processes the data inside the sensors instead of sending to the centralized servers, artificial intelligence analyzes the data based on pattern prediction, and the data is collected by the sensors.

There are multiple benefits of integrating edge intelligence in the EVs. The important benefits are as follows:

1. Real-Time Data Processing and Decision-Making: One of the primary advantages of edge intelligence in EVs is the ability to process data in real time. EVs generate vast amounts of data from various sensors, including cameras, radar, LiDAR, and other IoT devices. Processing these data at the edge allows for immediate responses to dynamic driving conditions, which is crucial for the safety and reliability of autonomous driving systems.
2. Enhanced Autonomous Driving Capabilities: Edge intelligence is critical for autonomous driving features. By processing sensor data locally, the vehicle can make split-second decisions, such as avoiding obstacles, changing lanes, or responding to traffic signals. This reduces latency compared to cloud-based processing and ensures that the vehicle can operate safely and efficiently even in areas with limited or unreliable internet connectivity.
3. Efficient Energy Management: EVs rely on batteries as their primary power source. Efficient energy management is vital to maximize the driving range and lifespan of the battery. Edge intelligence helps optimize energy consumption by analyzing driving patterns, environmental conditions, and battery health in real time. This enables predictive energy management strategies, such as adjusting regenerative braking or optimizing (Heating, Ventilation, and Air Conditioning) system usage, to conserve energy.
4. Predictive Maintenance and Reliability: Maintaining the health and reliability of EVs is essential for long-term performance. Edge intelligence enables predictive maintenance by continuously monitoring the vehicle's systems and components. By analyzing data from sensors and onboard diagnostics, the system can predict potential failures before they occur, allowing for proactive maintenance and reducing downtime.
5. Enhanced User Experience: Edge intelligence enhances the overall user experience by enabling personalized and responsive features. For example, infotainment systems can adapt to user preferences, and navigation systems can provide real-time traffic updates and route optimization. Additionally, edge intelligence can support advanced driver assistance systems (ADAS), offering features such as adaptive cruise control, lane-keeping assistance, and automated parking.
6. Security and Privacy: Processing data at the edge also improves security and privacy. Sensitive data can be analyzed and stored locally, reducing the risk of data breaches and unauthorized access. This is particularly important for autonomous vehicles, where the integrity and security of data are paramount.

Observing the importance of edge intelligence in EVs, we decided to provide the complete implementational details of edge intelligence in EVs with a case study. In the next subsection we describes the objectives of our considered case study.

### 13.1.2 Objectives of the case study

This case study is mainly focusing on the performance enhancement by introducing real-time data processing, dynamic resource allocation, and advanced user assistance systems increasing user experience in the EVs. The major objectives of proposing the case study are as follows:

1. To understand the importance of edge intelligence in the EVs;
2. To identify the key components in the EVs where edge intelligence can be implemented;
3. To figure out the challenges and issues present in the current implementations; and
4. To show the existing case studies of EVs and suggest future scope.

The rest of the chapter is organized as follows; Section 2 describes the literature study of edge intelligence implementation in EVs. Methodology used for the case study selection is shown in Section 3. Section 4 shows the description of the case study with its framework details. We discuss our case study's drawbacks and future research works in Section 6. Finally, we conclude in Section 7 with a summary.

## 13.2 Literature review

In this section, we first discuss the present day EVs' status and how edge intelligence is going to fulfill those requirements.

### 13.2.1 Current state of EVs

In the past five years, the use of EVs has been increasing because they reduce the $CO_2$ emission to the environment. Existing EVs are mainly divided into the following three categories: battery electric vehicles (BEVs), hybrid electric vehicles (HEVs), plug-in hybrid electric vehicles (PHEVs), and fuel cell electric vehicles (FCEVs). The BEVs are eco-friendly as they have zero mission but charging takes hours and finding a charging station is also difficult. Examples of BEVs are Tesla Model 3, Nissan Leaf, TATA Nexon, Mahindra E20 plus, MG ZS, Hyundai Kona, and Mahindra eVerito. HEVs have both engine and motor. The battery is the power source of the motor and fuel is the power source of the engine. In HEVs, charging cannot be done while plugged-in. Examples of HEVs are Chevrolet Tahoe Hybrid, Camry Hybrid, Lexus

NX, and Toyota Camry. PHEVs use an onboard charger and gasoline tank. PHEVs use smaller engines, so provide better efficiency; however, the running, maintenance, and replacement cost is very high. Examples of PHEV are Chevy Volt, Ford C-MAX Energi, and Hyundai Sonata. FCEVs are powered from hydrogen and emit only water vapor and warm air. The performance of FCEVs is very high but they depend on fossil fuels for hydrogen production. Examples of FCEVs are Toyota Mirai, Hyundai Nexo, Riversimple Rasa, and the Hyundai Tucson FCEV. The key trends perceived for EVs are as follows:

1. Market Growth: EV sales are going to be high (approx. 34 million) and with 100% $CO_2$ emission reduction by 2035. Public charging stations are increasing and fast charging is becoming more prevalent, especially in dense urban areas.
2. Battery Technology Advances: Lithium-ion battery costs have been declining, making EVs more affordable. Researchers are also working on solid-state batteries, which promise longer ranges and faster charging times. The average range of EVs has steadily increased, with many models now offering over 300 miles (500 km) on a single charge.
3. Charging Infrastructure Expansion: Countries are investing heavily in expanding public charging infrastructure, including fast-charging networks. In addition, many EV owners rely on home charging stations, and advancements in this area make it easier and faster to charge vehicles overnight.
4. Environmental and Policy Impacts: EVs are a central part of many countries' plans to reduce greenhouse gas emissions and combat climate change. Phasing out internal combustion engine (ICE) vehicles by certain deadlines (e.g., 2030–2040) is becoming a more common goal. Automakers are working to ensure that the entire EV supply chain, including battery production, is sustainable and ethical.
5. Autonomous and Connected EVs: EVs are increasingly becoming platforms for autonomous driving and connected car technologies. Many new EVs come equipped with advanced driver-assistance systems (ADAS) and full self-driving features are on the horizon.
6. Challenges: There are still challenges related to raw material sourcing, such as lithium, cobalt, and nickel, which are crucial for batteries. Ethical and environmental concerns are prompting efforts to find alternatives or improve extraction methods. The life cycle management of EV batteries is a growing area of focus, with increased efforts on recycling to minimize environmental impact.

### 13.2.2 Study of edge intelligence on EVs

Edge Intelligence (EI) is a paradigm that integrates artificial intelligence (AI) capabilities into edge computing, that means computing is done near the data sources (e.g., IoT devices, sensors) with optimization. In the context

*Table 13.1* Summary of Related Work on Edge Intelligence Implementation

| *Author* | *Contribution* |
|---|---|
| Wang et al. [3] | Highlighted comprehensive overview of the challenges and solutions in combining edge computing with AI, emphasizing distributed intelligence, real-time data processing, and new edge AI architectures. |
| Rodriguez et al. [4] | Provided a technical review of the existing frameworks that implement secure edge intelligence and various challenges related to secure AI deployment, and privacy-preserving edge intelligence. |
| Bao et al. [5] | Explored federated learning in the context of cloud-edge computing, addressing issues such as data heterogeneity, communication efficiency, and secure aggregation methods. |
| Stan et al. [6] | Focused on context-aware offloading decisions in edge intelligence, leveraging 5G and 6G capabilities and beyond for ultra-low-latency AI applications. |

of EVs, edge intelligence plays a crucial role in improving vehicle's performance, efficiency, and overall user experience, as the data processing needs to be done on-board and communicated to other devices as well as to the cloud. Edge intelligence has multiple key areas of research such as optimization of AI algorithms for edge devices, resource management, privacy and security, and use cases of the edge intelligence in smart cities, healthcare, and autonomous systems. In the abovementioned areas, only four research papers exist whose contribution summaries are presented in Table 13.1 Considering the literature and the research areas of edge intelligence in EVs, we can clearly state that no research paper exists that can highlight the research on edge intelligence implementation in EVs. Thus, we decided to provide details of a case study that includes edge intelligence in the EVs. In the next section, we explain the methodology for case study selection and framework analysis.

## 13.3 Methodology

In this section, we describe the methodologies used in the case study of edge intelligence.

### *13.3.1 Case study selection criteria*

While selecting case studies for developing edge intelligence in EVs it is essential to ensure that the selected cases provide valuable insights into the challenges and opportunities of integrating edge intelligence into EV systems.

The key criteria that can be considered when choosing case studies are as follows:

1. Relevance to Edge Intelligence in EVs:
   (i) Use of AI at the Edge: The case study must involve the application of AI or machine learning algorithms directly on the EVs or nearby edge devices rather than relying solely on cloud-based solutions. This includes tasks such as real-time data processing, predictive analytics, and decision-making.
   (ii) Edge Devices and Architecture: The case study should focus on the deployment of edge devices (e.g., onboard computers, sensors, or gateways) in the EV ecosystem for executing computational tasks.
2. Technological Maturity:
   (i) Advanced Technologies: Prioritize case studies that showcase advanced AI technologies such as deep learning, reinforcement learning, federated learning, or real-time analytics applied at the edge within the EV context.
   (ii) Integration with 5G Networks: The case studies must acknowledge the presence of 5G or next-gen network capabilities that support edge intelligence in EVs, such as low-latency communication between the vehicle and edge servers or infrastructure.
3. Application Scope:
   (i) Autonomous Driving: Case studies involving edge intelligence for autonomous driving or advanced driver assistance systems (ADAS) are crucial, as these represent high-demand applications for realtime decision-making at the edge.
   (ii) Energy Management and Efficiency: Cases where edge intelligence is used to optimize energy consumption, battery management, and charging infrastructure in EVs should be considered.
   (iii) Vehicle-to-Everything (V2X) Communication: Consider cases that involve edge intelligence for V2X, enabling the EV to communicate with other vehicles, infrastructure, and devices in their surroundings to enhance safety and efficiency.
4. Practical Outcomes:
   (i) Real-World Deployment: Prioritize cases that have gone beyond the simulation or experimental stages and are implemented in real-world environments, such as smart cities or autonomous fleets.
   (ii) Demonstrated Benefits: The case study should have clear metrics demonstrating the benefits of edge intelligence, such as reduced

latency, improved decision-making speed, enhanced safety, or optimized resource use in EVs.

5. Scalability and Replicability:

   (i) Scalability: The case study should focus on scalable solutions that can be applied to a wide range of EVs, including personal vehicles, commercial fleets, and public transportation systems.
   (ii) Replicability: The solutions or findings from the case study should be replicable across different EV platforms and geographic locations.

6. Challenges and Limitations:

   (i) Technical Challenges: The case study should explore challenges specific to edge intelligence in EVs, such as computational constraints, data security, model deployment, and real-time processing.
   (ii) Operational Constraints: Any limitations related to infrastructure, regulatory concerns, or technical bottlenecks that affect the implementation of edge intelligence in EVs should be addressed.

7. Collaboration and Ecosystem Development:

   (i) Industry–Academia Collaborations: Select case studies that involve collaboration between academic institutions, industry players, and governmental agencies, which can highlight multi-stakeholder efforts in developing edge intelligence solutions for EVs. (ii) Ecosystem Integration: The case should demonstrate how edge intelligence in EVs integrates with broader systems, such as smart grids, IoT networks, or smart city infrastructure.

8. Sustainability and Environmental Impact:

   (i) Energy Efficiency: Preference should be given to case studies that focus on improving the energy efficiency of EVs through edge intelligence, including reducing emissions, optimizing battery usage, and supporting renewable energy integration.
   (ii) Sustainability Goals: The case study should align with sustainability objectives, such as reducing the environmental footprint of transportation systems or contributing to cleaner, smarter cities.

9. Innovation and Novelty:

   (i) Innovative Solutions: Look for case studies that introduce novel edge intelligence frameworks or architectures, unique algorithms, or unconventional methods for overcoming technical challenges in EVs.
   (ii) Emerging Trends: Select cases that reflect emerging trends in edge intelligence for EVs, such as the use of blockchain for secure data sharing at the edge or quantum computing for enhanced optimization.

10. Geographical and Cultural Diversity:
    (i) Diverse Locations: Include case studies from various geographic regions to highlight differences in infrastructure, regulations, and consumer behavior that may impact the development of edge intelligence in EVs.
    (ii) Cultural and Regulatory Context: Consider how local regulations, cultural acceptance of autonomous technology, and regional EV adoption rates influence the deployment of edge intelligence.

### *13.3.2 Analysis framework*

The framework that needs to be used in the implementation of edge intelligence in EVs should include the evaluating technologies and outcomes considering the following objectives, characteristics, and challenges.

1. Objectives of Edge Intelligence Implementation: While analyzing the framework, the edge intelligence objectives needs to be clearly explained, such as the following:
    - Real-Time Processing: The need for on-the-spot data processing to reduce latency and enable real-time decision-making in critical scenarios.
    - Energy Efficiency: Evaluation of edge intelligence implementation to optimize energy consumption within the vehicle and the broader charging infrastructure.
    - Data Security and Privacy: How edge computing helps in maintaining data security by processing sensitive information locally rather than sending it to the cloud.
    - Resilience and Reliability: The goal of improving system resilience, especially in environments with intermittent connectivity.
2. Technical Framework and Architecture: The architecture of the model should be clearly explained considering the following key components:
    - Edge Computing Nodes: Examine the deployment of edge devices, such as onboard vehicle computers or charging station processors, and how they interact with the vehicle's sensors and control systems.
    - AI and Machine Learning Models: Look into the AI models deployed at the edge, focusing on their complexity, resource requirements, and ability to function under the constrained computational environment of edge devices.
    - Communication Infrastructure: Analyze the communication protocols and technologies used (e.g., 5G, V2X) that enable data exchange between the vehicle, edge nodes, and possibly the cloud.

- Data Management: Review how data are handled, stored, and processed at the edge, including considerations for data aggregation, filtering, and real-time analytics.

3. Implementation and Deployment: How the deployment needs to be done is another major component while analyzing the framework:

   - Pilot Deployment: Describe the initial deployment of the edge intelligence framework, including the selection of test environments (e.g., smart city infrastructure, highway systems).
   - Integration with Existing Systems: Evaluate how the edge intelligence solution was integrated with existing vehicle systems and infrastructure, such as traffic management systems or energy grids.

4. Testing and Validation: Look into the methodologies used for testing the edge intelligence system, including simulation environments, real-world trials, and performance metrics.
5. Challenges and Solutions: What are the challenges faced during the deployment and the probable solutions for the challenges:

   - Computational Constraints: Discuss the challenges related to the limited computational power of edge devices and how these were mitigated, for example, through model optimization techniques like pruning or quantization.
   - Latency Issues: Analyze the latency challenges in data processing and decision-making, and how edge computing solutions helped in reducing this latency.
   - Scalability: Consider how scalable the edge intelligence framework is, particularly in handling increased numbers of vehicles, charging stations, or more complex environments.
   - Security Concerns: Examine the security challenges posed by distributed data processing and how these were addressed through encryption, secure communication protocols, or federated learning.

6. Outcomes and Results: After the implementation, ascertain whether the outcomes match with the objectives or not.

   - Performance Metrics: Evaluate the success of the edge intelligence framework based on performance metrics such as reduced latency, improved energy efficiency, and enhanced real-time decision-making capabilities.
   - User Experience: Assess the impact of the edge intelligence system on user experience, particularly in terms of responsiveness, reliability, and overall vehicle performance.
   - Operational Efficiency: Analyze improvements in operational efficiency, including optimized energy use at charging stations, better traffic management, or enhanced vehicle maintenance schedules.

7. Lessons Learned and Best Practices: The technical insights and the collaborations need to be analyzed.
   - Technical Insights: Summarize the technical lessons learned from the implementation, such as the importance of low-latency communication or the need for lightweight AI models for edge devices.
   - Operational Considerations: Discuss the operational best practices identified during the project, such as strategies for deploying edge intelligence in varying environments or managing the balance between edge and cloud processing.
   - Collaboration and Integration: Highlight the importance of collaboration among stakeholders, including the integration of edge intelligence with existing infrastructure and systems.
8. Future Directions and Recommendations: Finally the future works for the same should be discussed.
   - Scaling the Framework: Discuss potential future developments, such as scaling the edge intelligence system to larger fleets of vehicles or more extensive charging networks.
   - Technological Advancements: Explore the role of emerging technologies, such as next-generation AI algorithms, enhanced 5G networks, or quantum computing, in further improving edge intelligence for EVs.
   - Policy and Regulation: Consider the role of policy and regulation in supporting the deployment of edge intelligence, particularly in areas like data privacy, cybersecurity, and standardization.

## 13.4 Case study description

In this section, we discuss as a case study the real-life edge intelligence implementation in EVs manufactured by Tesla. Before going to the case study directly, we mention the overview, goals, and the stakeholders, as follows:

### *13.4.1 Overview*

With the rise of EVs and autonomous driving technology, there is a growing need to process vast amounts of data in real time directly on the vehicle rather than relying solely on cloud-based systems. This is particularly crucial for EVs, which must make rapid decisions based on sensor inputs like cameras, LiDAR, and radar. However, real-time decision-making requires overcoming latency issues, ensuring data security, and optimizing energy consumption—all challenges that edge intelligence can address.

### *13.4.2 Goals of the case study*

The objective of this case study was to develop and implement edge intelligence solutions for EVs to enable real-time data processing, improve the accuracy and speed of decision-making, enhance safety, and reduce reliance on cloud-based infrastructure. The project was carried out as a collaboration between an EV manufacturer, Alliances, and technology firms specializing in AI.

### 13.4.3 Key stakeholders

The following are the major stakeholders in Tesla:

1. EV Manufacturer Company (EVMC): Provided the Autonomous Electrical Vehicles (AEVs) and expertise in vehicle systems. Although Tesla produces most of the parts for its cars internally, it purchases certain parts from outside vendors, such as brakes and windshields. These vendors are crucial collaborators that help Tesla manufacture and distribute its cars effectively.
2. AI Technology Firm (TechAI): Developed the AI algorithms and edge intelligence framework for real-time processing and decision-making. Tesla mostly designs all its AI algorithms itself for the onboard processors. In addition, Tesla has collaboration with NVIDIA for their AI processors implementation.
3. Alliances: In an effort to improve its standing in the EV industry, Tesla has forged strategic partnerships. To increase its market presence, it, for example, worked with Toyota to jointly develop electric car systems and parts. Furthermore, Tesla and Panasonic are building a manufacturing facility in New York with the goal of producing solar cells and batteries on a massive scale.
4. Government: The US federal government offers tax incentives to Tesla because it places a strong emphasis on the development of environmentally friendly vehicles and energy. These rewards help Tesla's expansion plans even more, as does the company's role in generating jobs in the United States.
5. Charging Points: Fast car charging stations are established by Tesla in partnership with a variety of establishments, including restaurants, retail centers, hotels, and resorts. Through these alliances, Tesla is able to increase the size of its charging network and provide its consumers more convenient ways to refuel.

### 13.4.4 Tesla and AEVs

Before providing detailed information regarding Tesla's AEV, we present one output example and then discuss the underlying architecture and framework. Figure 13.3 clearly shows one demo product of Tesla where the AI algorithms are implemented inside the onboard processors (Example of AI + Edge Intelligence).

**Technology behind the onboard processors:** Tesla has designed multiple AI algorithm–based chips, chip systems, and neural network–based vehicle

*Figure 13.3* Tesla's Trained EV [7].

infrastructures to integrate both AI and the edge intelligence. The board, algorithms, and the infrastructure details are as follows:

1. FSD Chip (Full Self-Driving Chip): FSD chips are used for energy efficiency in the EVs. In order to maximize silicon performance per watt, every tiny architectural and microarchitectural enhancement was taken into account during the design process. To confirm functionality and performance, they create robust testing, scoreboards, and floor plans, as well as timing and power evaluations for the design. Redundancy and energy optimization are the main topics of communication between the drivers program and the chip. Figure 13.4 represents the full self-driving chip's architecture.
2. Dojo Chip: The Dojo chips are the building blocks of the Dojo system. These are AI training chips to power Dojo systems. From the tiniest training nodes to the multi-die training tiles, the Dojo chip employs cutting-edge technology. At every level of detail, the architecture is designed for optimal speed, throughput, and bandwidth. The chip's physical design, floor plan, and other details are designed to guarantee functional accuracy, with techniques developed for postsilicon validation and pre-silicon verification. In order to maximize power and performance for neural networks across the board in the Dojo system, they also create compilers and drivers. Figure 13.5 shows an instance of a Dojo chip.

*Figure 13.4* Full Self-Driving Chip.

*Figure 13.5* Dojo Chip [8].

3. Dojo System: uses cutting-edge technology to solve complex problems in high-power supply and cooling, and creates scalable control loops and monitoring software. Utilizing the combined expertise of mechanical, thermal, and electrical engineering teams, Dojo system works on all facets of system design to develop the next generation of machine learning compute for Tesla data centers. It works along with Tesla fleet learning to develop a public facing API that will enable Dojo to be accessed by a large audience, and to deploy training workloads utilizing enormous datasets. Figure 13.6 shows the Dojo system's coupling of multiple Dojo chips.
4. Neural Networks: Neural networks used in Tesla's EVs utilize state-of-the-art research to train deep neural networks on perception and control challenges.

   The per-camera network module processes raw images to perform tasks such as object detection, monocular depth estimation, and semantic

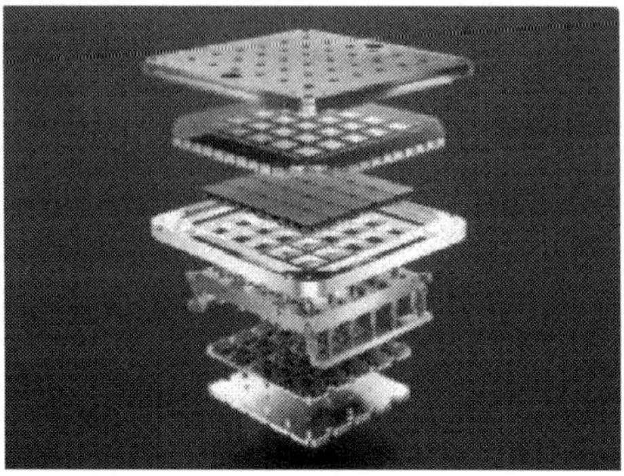

*Figure 13.6* Dojo Systems [9].

*Figure 13.7* Neural Networks.

segmentation. All camera footage is fed into the birds-eye-view networks, which output the top-down perspective of the road layout, static infrastructure, and three-dimensional objects. The networks that produce the top-down views of the road plans, stationary infrastructure, and three-dimensional objects are fed with all of the camera footage. It takes 70,000 GPU hours to train 48 networks in a full Autopilot neural network architecture. At every timestamp, they produce a total of 1,000 unique tensors, or forecasts. Figure 13.7 represents an instance of a neural network-enabled application.

5. Autonomy Algorithms: The fundamental algorithms that operate the car are developed by autonomy algorithms, which plot paths in the world and create a high-fidelity representation of it. Algorithms combine data from the vehicle's sensors in both space and time to provide large-scale, accurate ground truth data that may be used to train neural networks to predict such representations. They have developed a strong planning and

decision-making system that functions in challenging real-world circumstances with uncertainty by employing cutting-edge methodologies. The algorithms are assessed at the level of all Tesla vehicles. Figure 13.8 shows a working example of autonomy algorithms.

6. Code Foundations: The code foundations are basically optimized code primarily for throughput, latency, correctness, and determinism. They closely integrate with unique hardware, building the Autopilot software underpinnings from the bottom of the stack up. They have incredibly dependable bootloaders that can launch customized Linux kernels and allow over-the-air updates. To gather high-frequency, high-volume data from our sensors and share it with various consumer processes, quick, memory-efficient low-level codes are needed that do not affect core memory access latency or deprive crucial functional code of CPU cycles. Squeeze and pipeline computations are split among several system-on-chips and a range of hardware processor units. Figure 13.9 shows an example of code foundation.

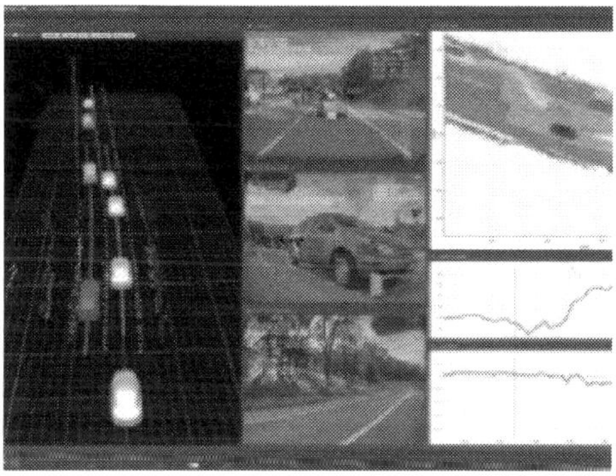

*Figure 13.8* Example of Autonomy Algorithms Implementation.

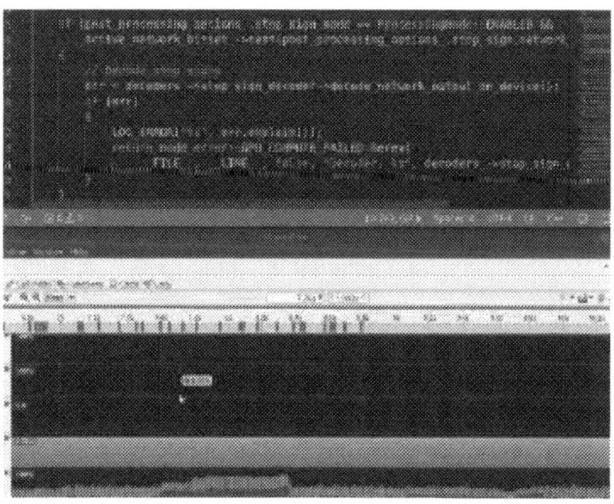

*Figure 13.9* Example of Code Foundations.

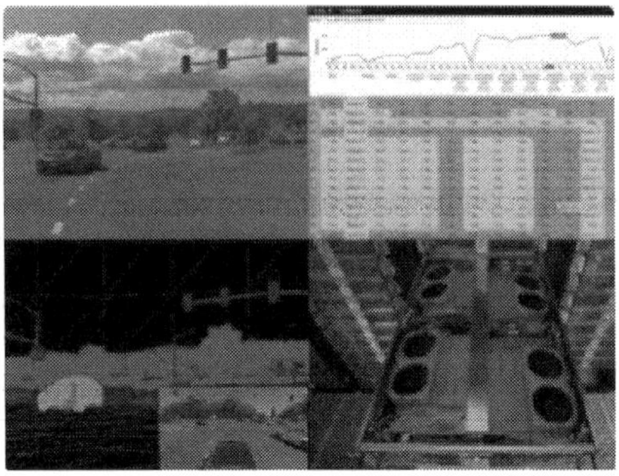

*Figure 13.10* Infrastructure Evaluation Example.

7. Evaluation Infrastructure: Building infrastructures and open and closed-loop hardware-in-the-loop assessment tools at scale are necessary to measure performance increases, stop regressions, and quicken the speed of innovation. In order to provide automated testing and real-time debugging, the infrastructure must supply highly realistic graphics and additional sensor data that feed Autopilot software, utilizing the fleet anonymized characteristic clips and merging them into large test case sets. Figure 13.10 represents an example of infrastructure evaluation in EVs.

## 13.5 How tesla's vehicles implement AI in the onboard processors?

Autonomous vehicles continuously analyze images from their sensors and machine vision cameras in order to drive themselves. They then utilize this knowledge to decide what to do next. They make use of AI to comprehend and predict the future motions of vehicles, pedestrians, and cyclists. They can quickly plan their moves and make split-second decisions thanks to this information. Should the vehicle change lanes or remain in its present lane? Is it better to pass the car ahead or hold its current position? When should the vehicle accelerate or brake?

To completely automate automobiles, Tesla has to gather the necessary data to feed their AIs and train their algorithms. Improved performance is a direct result of having more training data, and here is where Tesla shines.

Since all of the data from the hundreds of thousands of Tesla cars that are now on the road is crowdsourced, Tesla has a significant competitive edge. Both internal and external sensors track the actions of Teslas in a variety of scenarios. They also gather information on driver behavior, including how they respond to various circumstances and how frequently they touch the dashboard or steering wheel. The method used by Tesla is known as "imitation learning." Their algorithms pick up knowledge from the choices, responses, and maneuvers of millions of real drivers worldwide. Those many miles add up to extremely intelligent driverless vehicles. They have an extremely advanced tracking system. For instance, when a Tesla vehicle mispredicts how a car or bicycle will behave, Tesla records the event, adds it to the dataset, and then uses color-coded shapes to recreate an abstract picture of the situation that the neural network may learn from.

## 13.6 Discussions

Going through all the technologies involved during the implementation of AI in edge intelligence, we experienced certain challenges and arrived at future directions of research, which are as follows.

### *13.6.1 Lessons learned*

1. Battery: Battery performance of Tesla is largely reduced under cold environments. For example, all Tesla E-motors are powered by lithium batteries, a study conducted by the Shanghai Industrial Technology research institution showed that as the temperature decreases, Lithium-ion batteries' charging and discharging performance significantly degrades. At room temperature, the battery's discharge capacity is 87.0% of its total discharge capacity, while at—30 °C, its average discharge voltage drops by 0.598 V; Only 14% of the entire charging capacity of lithium-ion current can be charged at a constant current, and the charging time at a constant voltage increases [10, 11].
2. Competitors: The market share of Tesla is shrinking because of the increasing competition from traditional car manufacturers and new electric car companies. When it was first established, Tesla had less competition since traditional manufacturers were not as aggressive in entering the EV industry. The market for EVs is now being invaded by General Motors (GM), Ford (F), Volkswagen (VWAGY), and Toyota (TM), so that is no longer the case. On the other hand, in China NIO Inc., WuLong electric car company, and in India Electrical Appliance Manufacturing Companies are also joining the market of electric cars. These potential competitors would be threatening Tesla from aspects such as price and technology.
3. Charging Speed: Studies reveal that some of the objective factors limiting the development and practical application of the advantages of EV technology are its shortcomings such as the degradation of power characteristics of power batteries and the reduction in coulomb efficiency of charging and discharging caused by low temperatures. In contrast to conventional cars, which may be fully charged in a matter of seconds, electric cars typically require many hours to reach their maximum charging state. For instance, using a household electrical charger to fully charge a Tesla

Model 3 would take ten to twelve hours. Therefore, Tesla would need to find a solution to the issue of sluggish charging speed in both normal and cold situations if it wanted to stand out from the competition and fight for customers with traditional automakers.

### *13.6.2 Future research development*

First, the method of circulating high-temperature gas heating could be installed. This method refers to using air as a medium to directly pass through the power battery module to achieve the purpose of heating the power battery pack. Generally, forced air convection is used, which means that hot air is sent into the power battery box through an external fan or other devices to exchange heat with the power battery. Hot air can be generated by heating pads, and can also be obtained by using the heat emitted by the motor and high-power electronic and electrical heating devices in the vehicle. For hybrid vehicles, the engine can also provide energy to heat the air. This method could effectively heat up batteries to have them charging at their prime.

Second, to be price-appealing, Tesla needs to establish cheaper cars in order to compete with emerging electric car companies such as NIO from China.

Third, when electric cars are operating under low temperatures, the internal heating method could be used. The power battery is heated internally using the joule heat produced when an electric current flows through a conductor with a specific resistance value. In this case, the conductor is the actual power battery. Low temperatures cause the electrolyte inside the power battery to become more viscous, which restricts the movement of the charge carrier and raises the power battery's internal impedance. The electrolyte may even freeze under severe circumstances. However, taking advantage of the increased impedance of the power battery under low-temperature conditions, the impedance heat generation method can be used to maintain the operating temperature of the power battery. Pesaranl compared internal [2] heating with air heating and found that internal heating requires less energy and is more economical. According to the positive and negative flow directions of the current, it can be divided into the charging heating method, discharging heating method, and AC excitation heating method. According to the different power sources that provide the current, it can be divided into self-loss heating and external energy supply heating. Overall, by using the impedance heat generation method, Tesla as well as other electric cars could be operating properly under subzero harsh environments.

## 13.7 Conclusion

Through this chapter, we tried to first to introduce EVs' working principle, edge computing basics, and how edge intelligence is one of the major driving forces in the EVs (EVs). Second, we provided the technology requirement behind the integration of AI and edge computing in the EVs. As little research exists in the same area, we highlighted the challenges faced by the companies trying to implement edge intelligence in EVs. In addition, we suggested the future research directions to overcome the challenges faced by the industry. A complete case study on Tesla's implementation of edge intelligence in the EVs was discussed. In summary, we have given an overview of edge intelligence implementation in EVs using the technologies, criteria, and reallife examples, which will lead to develop basic understanding of edge intelligence in EVs and how to implement it.

## References


1. Program 18: Electric Transportation Overview. Available online: https://www.epri.com/research/programs/053122/overview (accessed on 5 July 2024).
2. Fayez Alanazi, "Electric Vehicles: Benefits, Challenges, and Potential Solutions for Widespread Adaptation", Applied Sciences Journal, MDPI, Vol. 23, 2023.
3. Xiaofei Wang, Yiwen Han, Victor C. M. Leung, Dusit Niyato, Xueqiang Yan, and Xu Chen, *Edge AI: Convergence of Edge Computing and Artificial Intelligence*, Springer, 2020.
4. E. Rodriguez, M. Pérez, A. Bastida, C. Senderos, and J. Armentia, "Edge Intelligence Secure Frameworks: Current State and Future Challenges", Computers & Security, Vol. 130, 2023.
5. Guanming Bao, and Ping Guo, "Federated Learning in Cloud-Edge Collaborative Architecture: Key Technologies, Applications and Challenges", Journal of Cloud Computing, Vol. 11, 2022.
6. C. Stan, S. Rommel, J. Olmos, and I. Monroy, "Computation Offloading in Beyond 5G/6G Networks with Edge Computing: Implications and Challenges", 20th International Conference on Distributed Computing and Artificial Intelligence (DCAI), 2023.
7. Tesla's Trained EV Example. Available online: https://bernardmarr.com/how-tesla-is-using-artificial-intelligence-to-create-the-autonomous-cars-of-the-future/ (accessed on 5 July 2024).
8. Tesla Dojo. Available online: https://en.wikipedia.org/wiki/Tesla_Dojo (accessed on 5 July 2024).
9. Dojo System. Available online: https://www.tomshardware.com/news/tesla-d1-ai-chip (accessed on 5 July 2024).
10. Sachin B. Chougule, Bharat S. Chaudhari, Sheetal N. Ghorpade, and Marco Zen-Naro, "Exploring Computing Paradigms for Electric Vehicles: From Cloud to Edge Intelligence, Challenges and Future Directions", World Electric Vehicle Journal, MDPI, Vol. 15, 2024.
11. Minzhe Li, "Analyze the Underlying Problems of Tesla Electric Cars and Other Similar Electric Cars", 7th International Conference on Economic Management and Green Development, 2023.

Chapter 14

# Scalability and reliability issues for edge intelligence

## Challenges and future directions

*Upendra, Rakesh Tripathi, Tirath Prasad Sahu, and Gurudatta Verma*

## 14.1 Introduction

### 14.1.1 Background and motivation

Edge intelligence is the conceptual framework that combines artificial intelligence (AI) with edge computing to achieve real-time processing of data near its source for making decisions or analytics. These applications include the Internet of Things (IoT), smart urban environment, and autonomous vehicles, all of which concern latency, privacy of data, and scalability in application [1]. Scalability is an issue because edge systems must process massive amounts of data as well as escalating user demands on a system to keep the same level of system performance. Similarly, reliability becomes crucial to have consistent and accurate operation under different and changing environments [2]. These issues are necessary to be addressed so that the real power of edge intelligence can be utilized to help latency-sensitive critical applications through resilient and scalable architectures. This chapter addresses the critical issues concerning scalability and reliability while providing some possible solutions and directions for future research.

### 14.1.2 Objectives of the chapter

The aim of this chapter is to explore the double challenge of scalability and reliability under the umbrella of edge intelligence: a domain at the intersection of AI with edge computing toward real-time data processing and decision-making. Scalability issues arise with the need for dealing with volumes of data, as well as user demands that grow exponentially while maintaining system efficiency and responsiveness. In such scenarios, reliability is crucial to ensure constant performance in dynamic and heterogeneous environments, especially in latency-constrained applications such as IoT and autonomous systems [1]. This chapter systematically reviews these challenges, evaluates state-of-the-art solutions such as resource optimization, federated learning, and fault-tolerant designs, and identifies research gaps. It has

DOI: 10.1201/9781003512066-14

taken these issues head-on in charting the roadmap for building scalable and reliable edge intelligence systems with the infusion of innovative technologies like AI-driven frameworks and blockchain to enhance the functionality of edges [2].

## 14.2 Overview of edge intelligence

### *14.2.1 Definition and core concepts*

Edge intelligence is a new paradigm that combines the proximity of decision-making and data processing with data generation near the source, through the combination of AI and edge computing. Edge intelligence reduces latency, enhances data privacy, and consumes less bandwidth than the traditional cloud-centric models [3]. Applications of edge intelligence are quite diversified and range from IoT devices in smart cities to autonomous vehicles and real-time monitoring systems. For example, in intelligent transportation systems, edge intelligence enables the processing of local traffic data to make timely and reliable decisions [4]. This fusion of AI with the edge makes possible heterogeneous devices interacting in collaboration while allowing smooth interplay and efficiency in decentralized spaces.

### *14.2.2 Current trends and technologies*

Recent technological advancements in the areas of fifth-generation (5G) networks, AI accelerators, and federated learning are profoundly changing the evolution of edge intelligence. The fast, low-latency features of 5G networks make it capable of allowing simultaneous and instant communication among devices, which are important for applications such as autonomous systems and remote healthcare. AI accelerators, which also include dedicated hardware for deep learning, advance the ability of the edges to manage computationally demanding tasks [5]. Federated learning has now become an important technology, making it possible for decentralized devices to collaboratively AI models while keeping raw data confidential and compliant with the regulatory standards. Such developments combine to enable edge intelligence to fulfill the needs of modern applications while at the same time tackling scalability and security issues.

## 14.3 Scalability challenges in edge intelligence

This chapter aims to explore the dual challenges of scalability and reliability in the context of edge intelligence, a field that integrates AI and edge computing to enable real-time data processing and decision-making, as shown in Figure 14.1.

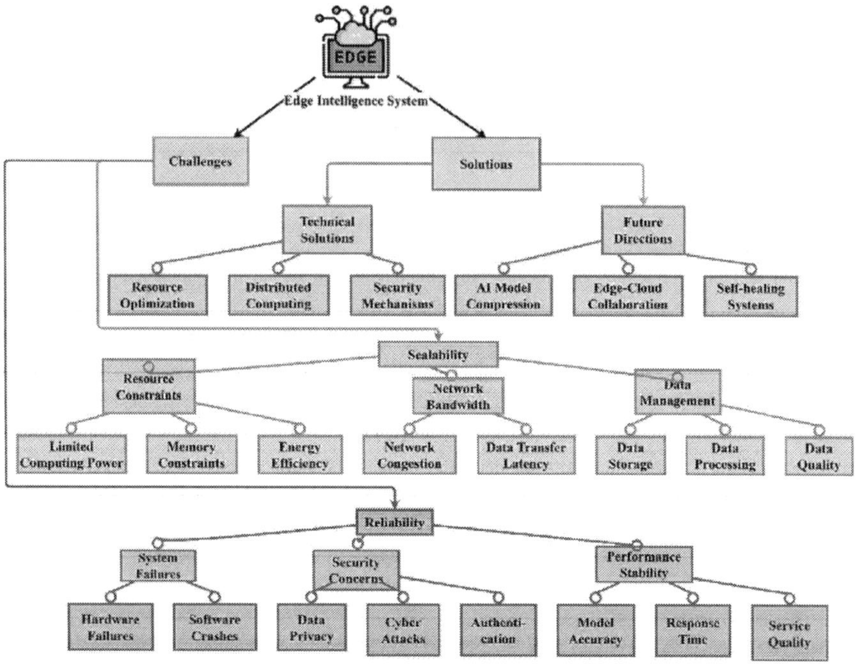


*Figure 14.1* Scalability and reliability challenges in edge intelligence, with proposed solutions and future directions.

### *14.3.1 Key scalability issues*

#### *14.3.1.1 Handling large-scale data in distributed edge systems*

The rapid deployment of IoT and edge-based applications has led to an unprecedented boom in data creation. Distributed edge systems need to process massive, heterogeneous data coming in real time, which further raises the question of scalability in such scenarios. Traditional centralized processing frameworks are highly unsuitable because of high latency, bandwidth bottlenecks, and security breaches [6]. To reduce reliance on cloud-based infrastructure, there is a need for edge intelligence to process and analyze data locally; however, the computational capability of the edge nodes limits large-scale data management. In addition, fluid data streams in distributed systems aggravate these challenges and therefore require adaptive mechanisms for efficient data processing.

#### *14.3.1.2 Resource allocation challenges*

Resource allocation in an edge computing setup is a way to achieve scalability. However, fundamental limitations of an edge node prove a challenge; these include limits on CPU abilities, memory available, and available energy, and this complicates the efficient spreading of workloads across multiple devices. Moreover, the changeable nature of workloads demands adaptive allocation techniques that adapt in real time to those changes [7]. Inefficient resource allocation can lead to bottlenecks, degraded performance, and increased latency, especially in large-scale IoT deployments or smart city infrastructures. Scalable edge systems require robust scheduling and task distribution mechanisms to address these challenges.

#### *14.3.1.3 Optimization challenges*

Optimization in edge computing systems should address performance, resource usage, and scalability. This can be very demanding because edge devices have to perform within strict energy boundaries while delivering the high level of performance needed to support real-time applications. Current optimization techniques, such as task offloading and data compression, solve some of the above problems but incur latency and computational overhead [8]. Future work should be directed toward developing energy-aware algorithms that make the most use of computation resources without sacrificing scalability. The integration of AI-based optimization techniques will further improve task scheduling and resource utilization, thus allowing for the development of more efficient distributed edge systems.

### *14.3.2 Examples of scalability bottlenecks*

#### *14.3.2.1 Real-world case 1: GPU-friendly subgraph isomorphism in large-scale graphs*

Scalability in edge intelligence is further challenged by computational demands when using large-scale data, especially in graph-based analytics. For example, subgraph isomorphism is one of the most critical tasks in graph databases, suffering from bottlenecks because of its intrinsic computational complexity. The further scale-up of traditional algorithms when analyzing larger graphs, especially with architectures using a GPU, is beyond the limitations. This proposed Scalable Graph Stream Ingestion algorithm can beat these limitations since it uses a vertex-centric framework with partition-based pipelines to facilitate the efficient processing of graphs containing billions of edges. This innovation points out some of the problems and potential means through which edge-based systems might be improved for processing large datasets [9].

#### *14.3.2.2 Real-world case 2: multi-exit DNN inference in edge computing*

Currently, edge intelligence systems consisting of DNNs suffer from bottlenecks arising from high transmission overhead as well as limited resources for computational hubs in the inference model. Multi-exit DNNs, which were designed to make possible early exits from inference tasks, pose significant challenges in optimizing their exit configurations, model partitioning, and resource allocation. MAMO is an acceleration framework for multi-exit DNNs that can overcome these challenges by using dynamic programming and deep reinforcement learning (DRL) to optimize the exit configuration and distribution of resources. This methodology reduces latency significantly, and therefore, allows edge systems to manage dynamic workloads much better [10].

#### *14.3.2.3 Real-world case 3: cost-effective strong consistency in geo-diverse edge systems*

Distributed systems suffer from strong inconsistencies and scalability bottlenecks. Considering distributed systems management, the Raft algorithm presents a clear example of performance penalties as the system scales across geo-diverse sites. BW-Raft takes on these bottlenecks by using lightweight abstractions that are stateless in nature, such as secretary and observer nodes, offloading intensive operations. Real-life experiments on cloud platforms portray the efficiency of BW-Raft, which achieves up to 9X output improvement for large-scale deployments, hence exhibiting its ability in mitigating the scalability issue within distributed edge systems [11].

### *14.3.3 Current solutions*

#### *14.3.3.1 Quality scalable quantization for deep learning models*

The scalability of edge intelligence is often hindered by the enormous resource requirements tied to deep learning models, specifically convolutional neural networks (CNNs). Parameter compression and adaptive multipliers offer Quality Scalable Quantization to resolve these issues by converting CNN filters into three-bit representations that drastically reduce both size and the computational demands on deep learning models for their practicality in edge devices. The experiments showed the potential for 82% savings in memory by retaining near-state-of-the-art accuracy and great promise for the deployment of resource-efficient deep learning on edge systems [12].

#### *14.3.3.2 Multi-exit DNN acceleration framework*

The Multi-Exit DNN Acceleration Framework is a new direction in solving the scalability problem that optimizes the inference operations of deep neural networks for edge intelligence. MAMO applies multidimensional optimizations, including exit selection, model partitioning, and resource allocation, to optimize latency and achieve efficiency. It has been proved that the edge computing environment that uses MAMO can deliver up to 13.7 times better

DNN inference results than the traditional approach. This framework integrates well with various tasks and varying network conditions, thus creating a robust solution for scalable edge systems [10].

#### *14.3.3.3 Federated learning with blockchain integration*

Federated learning (FL) proved a promising approach, addressing the main scalability challenges and edge systems along with decentralizing model training all while keeping information local. These blockchain-integrated FL frameworks, similar to FLCoin, exhibit enhanced scalability resulting from robust agreement protocols. The use of a double-layer blockchain structure reduces communication overheads while providing the maximum training efficiency, saving up to 90% communication cost and at least 35% training time. Such upgrades can make the FL systems more scalable and applicable over diverse IoT edge networks [13].

## 14.4 Reliability challenges in edge intelligence

### *14.4.1 Key reliability issues*

#### *14.4.1.1 Key issue 1: fault tolerance*

One of the major reliability challenges in edge computing systems is fault tolerance. Edge devices are working in different types of environments with resource-constrained scenarios that make them susceptible to hardware faults, network faults, and environmental stresses. Most of the conventional fault-tolerance strategies developed for cloud-based environments are resource-intensive and inappropriate for edge systems. New advances have brought forward resource-aware fault-tolerance strategies that combine fault tolerance with task placement using multiobjective optimization techniques. These techniques mitigate system malfunctions through the effective distribution of backup data, all while lessening resource expenditure [14].

#### *14.4.1.2 Key issue 2: data integrity*

This is a key issue because the edge devices are integrated with sensitive, sometimes real-time information, and data integrity must be maintained. The edge systems are highly prone to data corruption caused by hardware failure, software update, and network congestion. Sophisticated techniques for enhancing data resilience and privacy have been proposed. An example includes hierarchical cloud–edge collaborative fault-tolerant storage frameworks that enhance data resilience and privacy. In them, erasure coding and secure data transmission techniques are used, which reduces network traffic and storage demands considerably, thereby raising data reliability [15].

#### *14.4.1.3 Key issue 3: real-time processing reliability*

Real-time processing becomes one of the most stringent necessities of edge applications, especially in latency-sensitive domains such as smart grids and autonomous systems. Attainment of reliability in real time processing is highly difficult due to the unknown nature of workloads and resource availabilities. Intelligent proactive fault-tolerance strategies based on machine learning models such as RNNs predict potential faults and trigger preemptive resource allocation. Such an approach has shown significant improvements over traditional reactive methods in terms of system reliability and response times [16].

### 14.4.2 Examples of reliability gaps

#### *14.4.2.1 Scenario 1: failures in real-time DNN inference scheduling*

Dynamic environmental conditions and variability in resources are among the leading causes of reliability gaps in edge intelligence systems for real-time inference tasks. For example, the failures in mobile-edge computing systems deployed with DNNs result from uncertain ES availability, thereby interrupting inference processes and affecting the reliability of services mainly for latency-sensitive applications. Reliability-conscious task scheduling algorithms, such as RTBL, aim to avoid the problem of unpredicted server crashes by using past knowledge together with real-time experience. These techniques are, however, still prone to unexpected server failures and are unreliable [17].

#### *14.4.2.2 Scenario 2: hardware failures in edge-deployed CNNs*

Applications such as autonomous driving and object detection depend on CNNs that are deployed on edge devices. However, such devices are prone to hardware failures due to resource constraints and environmental factors, including temperature fluctuations or vibrations. Hardware failures may lead to failure in CNN computations and produce unreliable results. Traditional redundancy techniques are too resource-intensive to be used in edge systems. Very recent approaches, for example, ensemble-based reliability enhancement, try to address this through resource-efficient ensembles using pruned CNNs but still remain a challenge for robustness in extreme conditions [18].

#### *14.4.2.3 Scenario 3: reliability loss in fault-tolerant SoC systems*

In safety-critical applications, SoC designs include accelerators for AI customized for multiprecision DNNs. However, hardware-dependent issues cause variability in reliability because of neutron-induced soft errors affecting bit-level operations. Experimental research using fault injection techniques and radiation testing on SoC demonstrators revealed vulnerabilities in key components, including instruction memory and control status registers. The errors have a larger impact on the computations involving lower precision and further exacerbate the classification errors in the DNN. Although fault-tolerant designs have improved over time, maintaining such systems consistently with reliability is tough [19].

### *14.4.3 Existing mitigation approaches*

#### *14.4.3.1 Dynamic reliability management in multi-gateway edge systems*

In the multi-gateway edge computing scenario, dynamic reliability management (DRM techniques have been developed to enhance system resilience. Optimization of computation offloading and data routing is considered an important approach because it minimizes failure events and maximizes battery lifetime. In DRM architecture, controllers are multitiered with the feature of dynamic task assignment and bottlenecks during runtime. Based on the results of experiments taken using simulations, the designs show the ability to enhance Quality of Service (QoS) within reliability thresholds by several orders of magnitude [20].

#### *14.4.3.2 Concept drift-based checkpoint restart mechanisms*

The Concept Drift-Based Checkpoint Restart (CDCrest) framework employs the application of machine learning techniques to detect anomalies in real-time reliability data streams that are directly correlated with runtime reliability-related issues. It detects degradation of performance by environmental fluctuations with locality-sensitive hashing (LSH) and sliding window methodologies, which further enables the restarting of the checkpoints to restore the service functionality. This will ensure the continuation of edge services with reliability, even in unstable environments, and it is proved in experimental validations on actual datasets [21].

#### *14.4.3.3 Flexible shadow mechanisms*

The Flexible Shadow mechanism is a lightweight redundancy framework for enhancing the reliability of services at the often resource-constrained edge. Unlike traditional replication, where massive computational resources are consumed, the backup instance deployed by Flexible Shadow can adaptively adjust consumption based on demand. Experimental tests demonstrate that the method reduced computations by 60% while retaining subsecond latency, making it applicable in time-critical systems, such as autonomous systems, as well as use cases tied to the Industry 4.0 discourse [22].

## 14.5 Intersection of scalability and reliability

### 14.5.1 Interconnected nature

#### 14.5.1.1 Interconnected nature of scalability and reliability

Scalability and reliability are interdependent factors in edge intelligence architecture and functionality. Scalability allows the system to be capable of increasing its capacity when the workloads increase. Reliability guarantees consistent operation in different circumstances. The interplay of these two factors complicates the management of edge systems as they expand mainly due to limited resources and dynamic networks. Scalability in an edge system often means the distribution of workload among multiple nodes. However, when tasks are unevenly distributed across those nodes, they compromise reliability leading to possible node failures and service disruptions, therefore effective task orchestration to find a balance across these aspects is important [23].

#### 14.5.1.2 Trade-offs in system design

The scalability achieved sometimes involves trade-offs in the design, which affect reliability. For example, in a network with a high number of devices for enhanced scalability, synchronization and fault tolerance are added complexity issues. On the other hand, achieving redundancy to increase reliability would mean a heavy utilization of resources, which degrades system scalability. The latest frameworks such as DRM offer optimal computation offloading and resource allocation that would strike a balance between scalability and reliability. DRM techniques improve system reliability without impacting scalability, and hence they can be used for large-scale IoT deployments [20].

#### 14.5.1.3 Future perspectives

Innovative technologies like blockchain and AI-based optimization are significant in solving the confluence of scalability and reliability. The decentralized framework of blockchain enhances reliability by ensuring secure and consistent data management across multiple nodes, while AI algorithms enable predictive fault identification and adaptive resource distribution. These technologies help develop scalable and reliable edge systems that can accommodate a wide range of applications, including smart city infrastructure and industrial automation processes. For example, the use of blockchain-integrated edge frameworks has been shown to increase system reliability in handling large amounts of data [24].

### 14.5.2 Trade-offs in system design

#### 14.5.2.1 Trade-offs in resource allocation

In edge computing, achieving a good balance between scalability and reliability needs careful resource provisioning. As a corollary of that requirement, examples abound of the relation between scalability and reliability. Some DRM frameworks improve resource usage dynamically by reassigning tasks in real time to iron out bottlenecks in a multi-gateway system. There is, however, a cost: if one focuses on reliability, that means increased redundancy is likely and that could increase complexity in scheduling tasks. Studies have shown that DRM approaches can significantly enhance system reliability while maintaining acceptable scalability levels by using advanced optimization techniques like bottleneck assignment and hierarchical control [20].

#### 14.5.2.2 Latency and redundancy challenges

Latency and redundancy raise the complexity to a new dimension in balancing between scalability and reliability. One aspect of edge systems, Virtual Network Function (VNF) placement, is an area where this trade-off comes in. While the redundancy may achieve reliability through function replication across nodes, it results in increased latency and overall low scalability. Therefore, joint latency–reliability optimization algorithms have been proposed. These algorithms consider task dependencies and network constraints to minimize latency while maintaining reliable service delivery, but they often require significant computational resources, impacting scalability [25].

#### 14.5.2.3 Energy and scalability trade-offs

Energy consumption is one of the significant factors that balance scalability and reliability in edge computing systems. Systems designed for high reliability rely on the energy-intensive redundancy techniques, which can negatively affect scalability by providing fewer resources to additional tasks. Scalable reinforcement learning (RL) frameworks have been developed to overcome this problem without compromising the system's reliability. For example, a RL-based orchestrator for power management and retransmission in URLLC systems will potentially reduce energy consumption by 35% with reliability targets. However, such systems have to carefully manage resource availability to achieve effective scalability in large deployments [26].

## 14.6 Future directions

### 14.6.1 Emerging trends

#### 14.6.1.1 Artificial intelligence for enhanced scalability and reliability

AI is revolutionary in the area of edge computing with regard to offering intelligence for resource allocation, fault prediction, and task scheduling. Optimization frameworks based on AI like DRL, enable dynamic management of complex tasks by edge systems, thus being both scalable and dependable. For instance, AI-centric techniques have so far greatly shown improvements toward task offloading and collaborative caching that enhance real-time decision-making capabilities of the edge–cloud–terminal networks [27]. All these developments are geared toward increasing efficiency and reducing operational expenses, thus making AI a necessity for the future of edge intelligence.

#### 14.6.1.2 Blockchain for secure and decentralized systems

With the continued acknowledgment of blockchain technology as one of the prime enablers for secure, decentralized edge systems, such edge systems can, therefore, show improved data integrity and robust resilience against cyberattacks by integrating blockchain in edge intelligence. Decentralized data management solutions empower accountability and scalability in IoT-centric smart cities with the latest developments that have to do with Ethereum-based frameworks. More importantly, blockchain-based federated learning has been adopted to obtain high system reliability and privacy-preserving model training [28]. Such applications illustrate that blockchain is feasible to solve the twin problems of scalability and security.

#### 14.6.1.3 Quantum computing for advanced computational efficiency

Quantum computing is a disruptive advancement in solving computational bottlenecks in edge systems. Using quantum phenomena, such as superposition and entanglement, quantum algorithms can actually realize exponential improvement in complex operations like optimization and data analysis. The studies suggest that quantum computing may enhance applications of AI, enabling accelerated training and inference processes for massive edge networks. Although challenges will include hardware and the need for error correction, quantum computing is expected to significantly boost the scalability and reliability of future edge systems [29].

### 14.6.2 Research opportunities

#### *14.6.2.1 Enhancing resource allocation mechanisms*

Based on the discussion above, it can be noted that the scalability and reliability of edge intelligence systems are dramatically affected by the mechanisms of resource allocation. The conventional methodologies of allocation fail to adapt to dynamic and resource-constrained settings typical of edge systems. As such, future research efforts should be focused on designing adaptive frameworks for resource allocation based on real-time AI-driven optimization methodologies such as RL and federated learning. These models would ensure the best utilization of edge resources while preserving the reliability of service in very dynamic environments. For example, RL frameworks that use blockchain have shown efficacy in resource contention and general efficiency in a system [30].

In addition, the use of blockchain technology in the management of decentralized resources opens great opportunities. Combining blockchain's distributed ledger functionality with edge computing allows systems to achieve better fault tolerance and security, thereby ensuring scalability and reliability in distributed networks. Recent studies show that these hybrid approaches can be applied to overcome the problems of resource allocation and strengthen collaborative decision-making in edge settings [31].

#### *14.6.2.2 Leveraging emerging computing paradigms*

New paradigms of computing, including quantum computing and neuromorphic computing, provide new avenues for addressing scalability and reliability challenges. Quantum algorithms that take advantage of superposition and entanglement can speed up data processing and optimization tasks in edge systems. Quantum algorithms cannot, however, break through current computational bottlenecks. Neuromorphic computing based on the architecture of the human brain is energy-efficient for real-time processing, which is ideal for resource-constrained edge devices. Investigations may explore the integration of these paradigms into edge intelligence workflows to enhance both system efficiency and reliability.

#### *14.6.2.3 Advancing collaborative edge intelligence*

The cooperation between edge devices would enable distributed intelligence to greatly increase both scalability and reliability. Further research can be initiated into developing resilient frameworks that allow smooth collaboration between a wide range of devices. These frameworks have to address such problems as data synchronization, assignment of tasks, and fault tolerance. Moreover, federated learning integrated with edge computing opens ways to extend models of distributed AI while protecting the privacy of data. This approach is particularly promising for applications in smart cities and healthcare, where data security and real-time decision-making are critical [32].

### 14.6.3 *Practical recommendations*

#### *14.6.3.1 Framework-based deployment*

Structured frameworks like EXPRESS 2.0 help manage edge intelligence systems for scalability and reliability. Modules such as intelligent service collaboration, AI application management, and data security form part of EXPRESS 2.0, which enhances smooth operation within edge AIoT settings. For example, in last-mile delivery systems, it has been applied to manage complex workflows that include UAVs and UGVs. These frameworks improve scalability through support for extensive deployments and ensure reliability through resilient service management and fault-tolerant architecture [33].

#### *14.6.3.2 Optimization through resource-oriented AI*

Efficient deployment of edge AI systems needs to optimize the resource utilization especially in resource-constrained environments. New models like the Multi-Paradigm Deployment Model (MPDM) integrates multiple AI paradigms for scalable and reliable service orchestration. MPDM optimizes system accuracy and deployment cost using weighted metric-based heuristic algorithms that ensure efficient use of edge–cloud collaborative intelligence. Practical experiments have validated its efficacy in large-scale inference systems, and the importance of multi-paradigm deployment models in real-world scenarios has been highlighted [34].

#### *14.6.3.3 Federated learning for collaborative intelligence*

FedSwarm is an adaptive federated learning framework, which provides the mechanism to improve scalability and reliability in edge intelligence deployment. FedSwarm improves the model accuracy as well as energy efficiency by dynamic management of local updates and server synchronization. The method has brought significant improvements in real-world AIoT deployments with reduced energy consumption and robust performance across distributed edge servers [35].

## 14.7 Conclusion

### 14.7.1 *Recap of challenges*

The edge intelligence systems are associated with scalability and reliability challenges. Essentially, these issues come mainly from increasing complexity in the distributed systems, huge volumes of data, and resource-constrained environments. Scalability issues include managing large-scale data streams, optimal resource allocation, and dynamic workloads. In the same way, reliability issues relate to fault tolerance, data integrity, and real-time processing in diverse and unpredictable conditions. This is worse for applications requiring ultra-low latency such as autonomous vehicles and IoT networks [18].

### 14.7.2 Summary of solutions

Innovative solutions have been developed to handle such challenges, emphasizing dynamic resource management, advanced fault-tolerance mechanisms, and federated learning frameworks. Technologies such as AI-driven optimization, blockchain integration, and redundancy models like Flexible Shadow give strong techniques for balancing scalability with reliability. MPDM and federated machine learning frameworks have shown a great potential to handle real-time workloads efficiently while reducing resource contention and at the same time maintaining system reliability [33].

### 14.7.3 Future directions

In anticipation of breakthroughs, new technologies such as quantum computing, neuromorphic architectures, and blockchain-facilitated federated learning will bring enormous opportunities to advance edge intelligence. The efforts should be toward the development of energy-efficient and scalable frameworks integrating these technologies so that scalability and reliability are solved in a more systematic manner. Furthermore, across disciplines such as AI, telecommunications, and cybersecurity, collaborative effort will expedite the development of innovative solutions tailored to meet the needs of next-generation applications.

## References


1. Xu, D., Li, T., Li, Y., Su, X., Tarkoma, S., Jiang, T., . . . Hui, P. (2021). Edge intelligence: Empowering intelligence to the edge of network. *Proceedings of the IEEE*, *109*(11), 1778–1837.
2. Bourechak, A., Zedadra, O., Kouahla, M. N., Guerrieri, A., Seridi, H., & Fortino, G. (2023). At the confluence of artificial intelligence and edge computing in IoT-based applications: A review and new perspectives. *Sensors*, *23*(3), 1639.
3. Gokilakrishnan, G., Vigneshkumar, M., Dhanamurugan, A., Bhasha, A., Subbiah, R., & Anandakumar, H. (2023, March). A Review of Applications, Enabling Technologies, Growth Challenges and Solutions for IoT/IIoT. In *2023 9th International Conference on Advanced Computing and Communication Systems (ICACCS)* (Vol. 1, pp. 2241–2250). IEEE.
4. Xu, X., Huang, Q., Yin, X., Abbasi, M., Khosravi, M. R., & Qi, L. (2020). Intelligent offloading for collaborative smart city services in edge computing. *IEEE Internet of Things Journal*, *7*(9), 7919–7927.

5. Christou, A. G., Stergiou, C. L., Memos, V. A., Ishibashi, Y., & Psannis, K. E. (2023, September). Revolutionizing Connectivity: The Power of AI, IoT, and Edge Computing for Smart and Autonomous Systems. In *2023 6th World Symposium on Communication Engineering (WSCE)* (pp. 56–60). IEEE.
6. Murturi, I., Egyed, A., & Dustdar, S. (2022). Utilizing AI planning on the edge. *IEEE Internet Computing*, *26*(2), 28–35.
7. Manihar, S., Rehman, T. B., Patel, R., & Agrawal, S. (2020). Intelligent and scalable IoT edge-cloud system. *International Journal of Advanced Computer Science and Applications*, *11*(8).
8. Zhou, Y., Shi, Y., Zhou, H., Wang, J., Fu, L., & Yang, Y. (2023). Toward scalable wireless federated learning: Challenges and solutions. *IEEE Internet of Things Magazine*, *6*(4), 10–16.
9. Zeng, L., Zou, L., & Özsu, M. T. (2022). SGSI–a scalable GPU-friendly subgraph isomorphism algorithm. *IEEE Transactions on Knowledge and Data Engineering*, *35*(11), 11899–11916.
10. Dong, F., Wang, H., Shen, D., Huang, Z., He, Q., Zhang, J., . . . Zhang, T. (2022). Multi-exit DNN inference acceleration based on multi-dimensional optimization for edge intelligence. *IEEE Transactions on Mobile Computing*, *22*(9), 5389–5405.
11. Du, Y., Xu, Z., Zhang, K., Liu, J., Stewart, C., & Huang, J. (2022). Cost-effective strong consistency on scalable geo-diverse data replicas. *IEEE Transactions on Cloud Computing*, *11*(2), 1764–1776.
12. Khaliq, S. A., & Hafiz, R. (2024). Quality Scalable Quantization Methodology for Deep Learning on Edge. *arXiv preprint arXiv:2407.11260*.
13. Ren, S., Kim, E., & Lee, C. (2024). A scalable blockchain-enabled federated learning architecture for edge computing. *PLoS One*, *19*(8), e0308991.
14. Kozar, A., Del Monte, B., Zeuch, S., & Markl, V. (2024). Fault tolerance placement in the Internet of Things. *Proceedings of the ACM on Management of Data*, *2*(3), 1–29.
15. Chen, J., Wang, Y., Ye, M., & Jiang, Q. (2023). A secure cloud-edge collaborative fault-tolerant storage scheme and its data writing optimization. *IEEE Access*, *11*, 66506–66521.
16. Theodoropoulos, T., Violos, J., Tsanakas, S., Leivadeas, A., Tserpes, K., & Varvarigou, T. (2023). Intelligent Proactive Fault Tolerance at the Edge Through Resource Usage Prediction. *arXiv preprint arXiv:2302.05336*.
17. Ma, H., Li, R., Zhang, X., Zhou, Z., & Chen, X. (2023). Reliability-aware online scheduling for DNN inference tasks in mobile-edge computing. *IEEE Internet of Things Journal*, *10*(13), 11453–11464.
18. Gao, Z., Liu, S., Zhao, J., Wang, X., Wang, Y., & Han, Z. (2024). Ensemble-based reliability enhancement for edge-deployed CNNs in few-shot scenarios. *IEEE Transactions on Machine Learning in Communications and Networking*, 2, 1062–1075.
19. Cheng, Q., Huang, M., Man, C., Shen, A., Dai, L., Yu, H., & Hashimoto, M. (2023). Reliability exploration of system-on-chip with multi-bit-width accelerator for multi-precision deep neural networks. *IEEE Transactions on Circuits and Systems I: Regular Papers*, *70*(10), 3978–3991.
20. Ergun, K., Ayoub, R., Mercati, P., & Rosing, T. Š. (2022). Dynamic reliability management of multigateway IoT edge computing systems. *IEEE Internet of Things Journal*, *10*(5), 3864–3889.
21. Wang, L., Liu, J., & He, Q. (2022). Concept drift-based checkpoint-restart for edge services rejuvenation. *IEEE Transactions on Services Computing*, *16*(3), 1713–1725.

22. Yang, L., Zhou, A., Ma, X., Zhang, Y., Li, Y., & Wang, S. (2024, July). Flexible Shadow: Enhancing Service Reliability in Resource-Constrained Edge Computing. In *2024 IEEE International Conference on Web Services (ICWS)* (pp. 767-777). IEEE.
23. Hossfeld, T., Heegaard, P. E., & Kellerer, W. (2023). Comparing the scalability of communication networks and systems. *IEEE Access*, *11*, 101474–101497.
24. Thakur, P., & Sehgal, V. K. (2024). Synergizing edge computing and blockchain for cyber-physical systems. *Concurrency and Computation: Practice and Experience*, *36*(12), e8066.
25. Haim, R. B., & Rottenstreich, O. (2023). Low-Latency and Reliable Virtual Network Function Placement in Edge Clouds. *IEEE Transactions on Network and Service Management*, *20*, 2172–2185.
26. Ganjalizadeh, M., Ghadikolaei, H. S., Azari, A., Alabbasi, A., & Petrova, M. (2023). Saving energy and spectrum in enabling URLLC services: A scalable RL solution. *IEEE Transactions on Industrial Informatics*, *19*(10), 10265–10276.
27. Gu, H., Zhao, L., Han, Z., Zheng, G., & Song, S. (2023). AI-enhanced cloud-edge-terminal collaborative network: Survey, applications, and future directions. *IEEE Communications Surveys & Tutorials*, 26(02), 1322–1385.
28. Khan, B. U. I., Goh, K. W., Khan, A. R., Zuhairi, M. F., & Chaimanee, M. (2024). Integrating AI and Blockchain for enhanced data security in IoT-driven smart cities. *Processes*, *12*(9).
29. Nagaraj, G., Upadhayaya, N., Matroud, A., Sabitha, N., Reshma, V. K., & Nagaraju, R. (2023, March). A Detailed Investigation on Potential Impact of Quantum Computing on Improving Artificial Intelligence. In *2023 International Conference on Innovative Data Communication Technologies and Application (ICIDCA)* (pp. 447–452). IEEE.
30. Hao, X., Yeoh, P. L., She, C., Vucetic, B., & Li, Y. (2023). Secure deep reinforcement learning for dynamic resource allocation in wireless MEC networks. *IEEE Transactions on Communications*, *72*(03), 1414–1427.
31. El-Taie, M., & Kraidi, A. Y. (2024). Crafting resilient consensus mechanisms for the web 3.0 network through edge intelligence. *International Journal of Wireless & Ad Hoc Communication*, *8*(1).
32. Meng, X. (2024). Optimization of algorithmic efficiency in AI: Addressing computational complexity and scalability challenges. *Applied and Computational Engineering*, *45*, 305–311.2222.
33. Xu, J., Liu, X., Pan, W., Li, X., Yao, A., & Yang, Y. (2023, September). EXPRESS 2.0: An Intelligent Service Management Framework for AIoT Systems in the Edge. In *2023 38th IEEE/ACM International Conference on Automated Software Engineering (ASE)* (pp. 2022–2025). IEEE.
34. Wang, L., Ren, X., Zhao, C., Zhao, F., & Yang, S. (2022). MPDM: A multi-paradigm deployment model for large-scale edge-cloud intelligence. *IEEE Internet of Things Journal*, *10*(10), 8773–8785.
35. Du, H., Ni, C., Cheng, C., Xiang, Q., Chen, X., & Liu, X. (2023). FedSwarm: An adaptive federated learning framework for scalable AIoT. *IEEE Internet of Things Journal*, *11*(05), 8268–8287.

# Index

Page numbers in *italic* indicate a figure and page numbers in **bold** indicate a table on the corresponding page.

Made in the USA
Las Vegas, NV
19 September 2026

86b6c764-b53e-4a4e-9197-2f9f29edcb20R01